Ultrasonic Techniques for Liquids Characterisation

Online at: https://doi.org/10.1088/978-0-7503-3203-3

Ultrasonic Techniques for Liquids Characterisation

Megan J Povey
School of Food Science and Nutrition, University of Leeds, Leeds, UK

With contributions from:
Valerie J Pinfield
Chemical Engineering, Loughborough University, Loughborough, UK

Daniel I Hefft
Chemical Engineering, University of Birmingham, Birmingham, UK

Alexander L Bowler
School of Food Science and Nutrition, University of Leeds, Leeds, UK

Nicholas Watson
School of Food Science and Nutrition, University of Leeds, Leeds, UK

IOP Publishing, Bristol, UK

ISBN 978-0-7503-3203-3 (ebook)
ISBN 978-0-7503-3201-9 (print)
ISBN 978-0-7503-3204-0 (myPrint)
ISBN 978-0-7503-3202-6 (mobi)

DOI 10.1088/978-0-7503-3203-3

Version: 20251201

IOP ebooks

British Library Cataloguing-in-Publication Data: A catalogue record for this book is available from the British Library.

Published by IOP Publishing, wholly owned by The Institute of Physics, London

IOP Publishing, No.2 The Distillery, Glassfields, Avon Street, Bristol, BS2 0GR, UK

US Office: IOP Publishing, Inc., 190 North Independence Mall West, Suite 601, Philadelphia, PA 19106, USA

Contents

Preface

A great deal of progress has been made since the publication of *Ultrasound Techniques for Fluids Characterisation* in 1997. It was written as a manual for those interested in characterising fluids using ultrasound techniques. When it was written, there was no commercially available ultrasound attenuation spectrometer, and it was far easier to make velocity of sound measurements than attenuation measurements. Machine learning (ML) and artificial intelligence (AI) are now a commonplace component in an engineer's toolbox, transforming non-invasive and 100% inspection techniques. The spectacular advance in available computing power has made high speed data acquisition, storage, and dissemination economically viable. Significant advances have been made in acoustical theory applied to materials characterisation. The authors of this book have all played a role in advancing ultrasound measurement as a practical tool in process analytical technology (PAT), high throughput sensing (HTS), materials characterisation, and process control. Each chapter is authored by experts in the field.

Ultrasonic techniques for liquids characterisation

Ultrasound scans of animal-based and plant-based cheese.

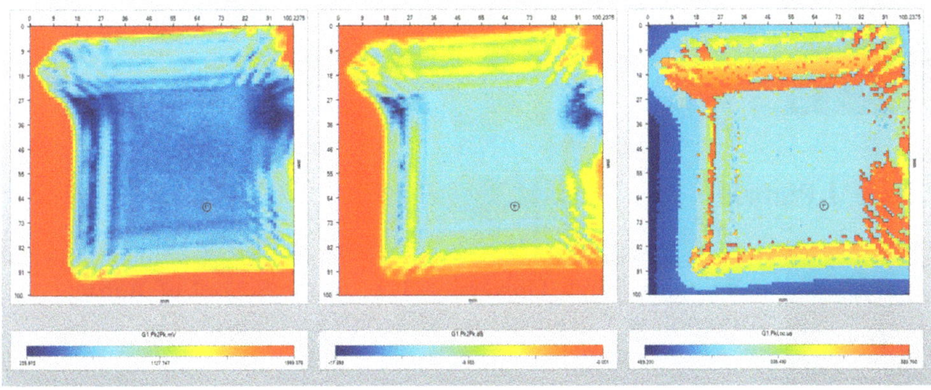

Author biographies

Megan J Povey

Megan J Povey is Professor of Food Physics in the School of Food Science and Nutrition at the University of Leeds.

Valerie J Pinfield

Valerie J Pinfield is Professor of Ultrasonics and Complex Materials at Loughborough University, UK, in the School of Aeronautical, Automotive, Chemical and Materials Engineering.

Daniel I Hefft

Dr Daniel I Hefft is R&D Process Engineering Manager at Medik8, advancing applied research at the intersection of physics, engineering, and material science. He is Honorary Research Fellow at the University of Birmingham School of Chemical Engineering.

Alexander L Bowler

Alexander L Bowler is a Research Fellow in Artificial Intelligence for Sustainable Food in the School of Food Science and Nutrition at the University of Leeds.

Nicholas Watson

Nicholas Watson is a Professor of Artificial Intelligence in Food in the School of Food Science and Nutrition at the University of Leeds.

List of contributors

Alexander L Bowler
School of Food Science and Nutrition, University of Leeds, Leeds, UK

Daniel I Hefft
Chemical Engineering, University of Birmingham, Birmingham, UK

Valerie J Pinfield
Chemical Engineering, Loughborough University, Loughborough, UK

Megan J Povey
School of Food Science and Nutrition, University of Leeds, Leeds, UK

Nicolas Watson
School of Food Science and Nutrition, University of Leeds, Leeds, UK

IOP Publishing

Ultrasonic Techniques for Liquids Characterisation

Megan J Povey

Chapter 1

Introduction

Ultrasound Techniques for Fluids Characterisation (Academic Press, 1997) [1] lays out the basic theoretical framework and the contemporary experimental techniques underpinning fluids characterisation by ultrasound. This current book presumes a familiarity with that work (which is still in print), particularly the first two chapters.

Since 1996, when *Ultrasound Techniques for Fluids Characterisation* was written, the author and her research group have published extensively on *crystallisation processes* [2–27], *buffer rod techniques* [4, 5, 16, 28, 29], *particle sizing, interactions and aggregation* [8, 10, 11, 13–17, 19, 22–24, 30–56], [52], *milk and milk fat characterisation* [2, 16, 30, 32, 38, 48], *crystallisation control* [3, 57–59], *machine learning and ultrasound* [4, 60], *fluids and soft solids characterisation* [2, 19, 23, 52, 61–70], *powder characterisation* [16, 30, 71], *acoustic diffractometry* [61, 72], *and non-invasive acoustic rheology* [2, 23, 52, 73–77].

The current book title refers to liquids rather than fluids since anything that is capable of flow, including soft solids, is a fluid.

The book updates the theoretical framework outlined in the 1997 work (chapter 2 Valerie Pinfield). This chapter assumes familiarity with mathematics, particularly vector theory. The developed theory is then applied in chapter 3 (Megan Povey and Daniel Hefft) to demonstrate the link between rheology and acoustic/ultrasound propagation. Chapter 3 and subsequent chapters do not assume the same level of mathematical competence. In chapter 4 (Alex Bowler and Nick Watson), contemporary experimental, process analytical (PAT) and high throughput sensing (HTS) methods are detailed and their enhancement through machine learning (ML) and artificial intelligence (AI) are described. In chapter 5, we provide examples of practical ultrasound and acoustic techniques and consider the integration of ultrasound technology into multi-modal measurement frameworks and tomographic applications. In the introduction to chapter 5 we consider the underlying physical and physico-chemical properties that the different modalities respond to.

doi:10.1088/978-0-7503-3203-3ch1

References

[1] Povey M J W 1997 *Ultrasonic Techniques for Fluids Characterization* (San Diego, CA: Academic) p 214

[2] Povey M J and Hefft D I 2023 Characterising the mechanical properties of soft solids through acoustics and rheology, exemplified by anhydrous milk fat *Soft Matter* **19** 8349–59

[3] Povey M J, Ettelaie R, Lewtas K, Price A, Lai X and Sheng F 2023 'Sounding' out crystal nuclei—a mathematical-physical and experimental investigation *J. Chem. Phys.* **158** 174501-1-11

[4] Metilli L, Morris L, Lazidis A, Marty-Terrade S, Holmes M, Povey M and Simone E 2022 Real-time monitoring of fat crystallization using pulsed acoustic spectroscopy and supervised machine learning *J. Food Eng.* **335** 111192

[5] Morris L, Simone E, Glover Z J, Powell H, Francis M, Povey M J *et al* 2021 Dynamic monitoring of glycine crystallisation with low power ultrasound reflection spectroscopy *Chem. Eng. Res. Des.* **0** 213–23

[6] Ladd-Parada M, Povey M J, Vieira J and Ries M E 2019 Fast field cycling NMR relaxometry studies of molten and cooled cocoa butter *Mol. Phys.* **18** 1020–7

[7] Ladd Parada M, Sadeghpour A, Vieira J, Povey M and Rappolt M 2018 Global small-angle x-ray scattering data analysis of triacylglycerols in the α-phase (Part II) *J. Phys. Chem. B.* **122** 10330–6

[8] Povey M J W 2017 Applications of ultrasonics in food science—novel control of fat crystallization and structuring *Curr. Opin. Colloid Interface Sci.* **28** 1–6

[9] Povey M J W 2016 Nucleation in food colloids *J. Chem. Phys.* **145** 211906

[10] Gregersen S B, Povey M J W, Andersen M D, Hammershøj M, Rappolt M, Sadeghpour A and Wiking L 2016 Acoustic properties of crystallized fat: relation between polymorphic form, microstructure, fracturing behavior, and sound intensity *Eur. J. Lipid Sci. Technol.* **118** 1257–70

[11] Povey M J W 2014 Crystal nucleation in food colloids *Food Hydrocoll.* **42** 118–29

[12] Parker N G, Mather M L, Morgan S P and Povey M J W 2010 Longitudinal acoustic properties of poly(lactic acid) and poly(lactic-co-glycolic acid) *Biomed. Mater.* **5** 1–10

[13] Povey M J W and Challis R E 2009 Applications of ultrasound to analysis/quantitation of dairy lipids *Adv. Dairy Chem.* **2** 709–23

[14] Povey M J W, Awad T S, Huo R and Ding Y L 2009 Quasi-isothermal crystallisation kinetics, non-classical nucleation and surfactant-dependent crystallisation of emulsions *Eur. J. Lipid Sci. Technol.* **111** 236–42

[15] Povey M J W, Hindle S A A, Aarflot A and Hoiland H 2006 Melting point depression of the surface layer in n-alkane emulsions and its implications for fat destabilization in ice cream *Cryst. Growth Des.* **6** 297–301

[16] Meyer S, Rajendram V S and Povey M J W 2006 Characterisation of reconstituted milk powder by ultrasound spectroscopy *J. Food Qual.* **29** 405–18

[17] Challis R E, Povey M J W, Mather M L and Holmes A K 2005 Ultrasound techniques for characterizing colloidal dispersions *Rep. Prog. Phys.* **68** 1541–637

[18] Mykhaylyk O O, Castelletto V, Hamley I W and Povey M J W 2004 Structure and transformation of low-temperature phases of 1,3-distearoyl-2-oleoyl glycerol *Eur. J. Lipid Sci. Technol.* **106** 319–24

[19] Oxtoby D W, Cahn R W, Povey M J W and Perepezko J H 2003 Crystal nucleation in simple and complex fluids *Philos. Trans. R. Soc. A.* **361** 419–28

[20] Muthukumar M, Povey M J W, Edwards S and Windle A H 2003 Molecular modelling of nucleation in polymers *Philos. Trans. R. Soc. A.* **361** 539–56

[21] Chow R, Blindt R, Chivers R and Povey M 2003 The sonocrystallisation of ice in sucrose solutions: primary and secondary nucleation *Ultrasonics* **41** 595–604

[22] Hindle S, Povey M and Smith K 2002 Characterizing cocoa-butter seed crystals by the oil-in-water emulsion crystallization method *J. Am. Oil Chem. Soc.* **79** 993–1002

[23] Nelson P V, Povey M J W and Wang Y 2001 An ultrasound velocity and attenuation scanner for viewing the temporal evolution of a dispersed phase in fluids *Rev. Sci. Instrum.* **72** 4234

[24] Hindle S, Povey M J W and Smith K 2000 Kinetics of crystallization in n-hexadecane and cocoa butter oil-in-water emulsions accounting for droplet collision-mediated nucleation *J. Colloid Interface Sci.* **232** 370–80

[25] Smith P R and Povey M J W 1997 The effect of partial glycerides on trilaurin crystallization *J. Am. Oil Chem. Soc.* **74** 169–71

[26] Dickinson E, Ma J and Povey M J W 1996 Crystallization kinetics in oil-in-water emulsions containing a mixture of solid and liquid droplets *J. Chem. Soc. Faraday Trans.* **92** 1213–5

[27] Sadeghpour A, Parada M L, Vieira J, Povey M and Rappolt M 2018 Global small-angle x-ray scattering data analysis of triacylglycerols in the molten state (Part I) *J. Phys. Chem. B.* **122** 10320–9

[28] Tong H and Povey M J W 2002 Pulse echo comparison method with FSUPER to measure velocity dispersion in n-tetradecane in water emulsions *Ultrasonics* **40** 37–41

[29] Povey M J W, Watson N, Parker N G, Morris V and Groves K 2013 *Ultrasonic and Acoustic Microscopy: Principles and Applications to Food Microstructures* (Woodhead Publishing) pp 192–222

[30] Glover Z J, Francis M J, Fiutowski J, Sun Q, Yu Q, Andersen U *et al* 2020 Acoustic attenuation spectroscopy and helium ion microscopy study of rehydration of dairy powder *Colloids Surf., A.* **600** 124795

[31] Francis M J, Glover Z J, Yu Q, Povey M J and Holmes M J 2019 Acoustic characterisation of pH dependant reversible micellar casein aggregation *Colloids Surf. A.* **568** 259–65

[32] Simonsen A C, Andersen U, Glover Z J, Povey M J, Brewer J R, Holmes M J *et al* 2018 Super-resolution microscopy and empirically validated autocorrelation image analysis discriminates microstructures of dairy derived gels *Food Hydrocoll.* **90** 62–71

[33] Svalova A, Parker N G, Povey M J W W and Abbott G D 2017 Determination of asphaltene critical nanoaggregate concentration region using ultrasound velocity measurements *Sci. Rep.* **7** 1–11

[34] Zou X, Huang X and Povey M 2016 Non-invasive sensing for food reassurance *Analyst.* **141** 1587–610

[35] Watson N, Hazlehurst T, Povey M, Vieira J, Sundara R and Sandoz J P 2014 Can airborne ultrasound monitor bubble size in chocolate? *J. Phys. Conf. Ser.* **498** 012001

[36] Hazlehurst T A, Harlen O G, Holmes M J and Povey M J W 2014 Multiple scattering in dispersions, for long wavelength thermoacoustic solutions *J. Phys. Conf. Ser.* **498** 012005

[37] Zhang L L, Li Y, Liu X M, Zhao L H, Ding Y L, Povey M and Cang D Q 2013 The properties of ZnO nanofluids and the role of H2O2 in the disinfection activity against *Escherichia coli Water Res.* **47** 4013–21

[38] Povey M J W 2013 Ultrasound particle sizing: a review *Particuology* **11** 135–47

[39] Hazlehurst T A, Harlen O G, Holmes M J and Povey M J W Iop 2014 Multiple scattering in dispersions, for long wavelength thermoacoustic solutions *12th Annual Meeting on Anglo-French Physical Acoustics (AFPAC) 2013 2014 Jan 16–18 (Frejus, France)*
[40] Luo Z, Murray B S, Ross A L, Povey M J W, Morgan M R A and Day A J 2012 Effects of pH on the ability of flavonoids to act as Pickering emulsion stabilizers *Colloids Surf. B.* **92** 84–90
[41] Abbott G and Povey M 2012 The acoustic spectroscopy of asphaltene aggregation in petroleum *International Symposium on Ultrasound in the Control of Industrial Processes (UCIP)* 2012**42**
[42] Luo Z, Murray B S, Yusoff A, Morgan M R A, Povey M J W and Day A J 2011 Particle-stabilizing effects of flavonoids at the oil-water interface *J. Agric. Food Chem.* **59** 2636–45
[43] Chen J S, Gaikwad V, Holmes M, Murray B, Povey M, Wang Y and Zhang Y 2011 Development of a simple model device for *in vitro* gastric digestion investigation *Food Funct.* **2** 174–82
[44] Zhang L L, Ding Y L, Povey M and York D 2008 ZnO nanofluids—a potential antibacterial agent *Prog. Nat. Sci.* **18** 939–44
[45] Zhang L L, Jiang Y H, Ding Y L, Povey M and York D 2007 Investigation into the antibacterial behaviour of suspensions of ZnO nanoparticles (ZnO nanofluids) *J. Nanopart. Res.* **9** 479–89
[46] Povey M J W 2006 Acoustic methods for particle characterisation *KONA Powder Part. J.* **24** 126–33
[47] Pinfield V J, Harlen O G, Povey M J W and Sleeman B D 2006 Acoustic propagation in dispersions in the long wavelength limit *SIAM J. Appl. Math.* **66** 489–509
[48] Meyer S, Berrut S, Goodenough T J, Rajendram V S, Pinfield V J and Povey M J W 2006 A comparative study of ultrasound and laser light diffraction techniques for particle size determination in dairy beverages *Meas. Sci. Technol.* **17** 289–97
[49] Pinfield V J, Harlen O G, Povey M J W and Sleeman B D 2005 Acoustic propagation in dispersions in the long wavelength limit *SIAM J. Appl. Math.* **66** 489–509
[50] Gregory M, Jones F, Malcolm J W, Povey H and John Campbell L Inventors 2005 Measurement and control of asphaltene agglomeration in hydrocarbon liquids *UK Patent* EP 1021710B1
[51] Harlen O G, Holmes M J, Povey M J W, Qiu Y and Sleeman B D 2003 A low frequency potential scattering description of acoustic propagation in dispersions *SIAM J. Appl. Math.* **61** 1906–31
[52] Povey M J W, Nelson P V, Tong J and Jones G M Inventors 2002 Baker Hughes, Inc., Assignee Particle measurement by acoustic speckle *Patent* EP1092976A2
[53] Povey M J W 2000 Particulate characterization by ultrasound *Pharmaceut. Sci. Technol. Today* **3** 373–80
[54] Wang Y and Povey M J W 1999 A simple and rapid method for the determination of particle size in emulsions from ultrasound data *Colloids Surf. B.* **12** 417–27
[55] Povey M J W, Golding M, Higgs D and Wang Y T 1999 Ultrasonic spectroscopy studies of casein in water *Int. Dairy J.* **9** 299–303
[56] Abbott G D and Povey M J W 2012 The acoustic spectroscopy of asphaltene aggregation in petroleum *IOP Conf. Ser.: Mater. Sci. Eng.* **42** 012022
[57] Povey M and Lewtas K Inventors 2022 Use of ultrasound and acoustics to control crystallisation *Patent US* 2022/0143528 A

[58] Lewtas K and Povey M J Inventors 2022 Use of ultrasound and acoustics to control crystallisation *Patent* GB 2548177 B
[59] Povey M J W and Lewtas K Inventors 2017 Poperty Organsiation, Assignee Use of ultrasound and acoustics to control crystallisation
[60] Zou X B, Zhao J W, Povey M J W, Holmes M and Mao H P 2010 Variables selection methods in near-infrared spectroscopy *Anal. Chim. Acta.* **667** 14–32
[61] Francis M J, Holmes M J, Povey M J, Pinfield V J, Gower A L, Huang J *et al* 2018 Characterisation of complex fluid media by an angle-scanning ultrasound diffractometer *J. Acoust. Soc. Am.* **144** 1758
[62] Ghosh S, Holmes M and Povey M 2017 Temperature dependence of bulk viscosity in edible oils using acoustic spectroscopy *J. Food Process. Technol.* **08** 1000676
[63] Parker N G, Mather M L, Morgan S P and Povey M J W 2011 A preliminary study of acoustic propagation in thick foam tissue scaffolds composed of poly(lactic-co-glycolic acid) *J. Phys. Conf. Ser.* **269** 012019
[64] Huynh N T, Parker N G, He D, Ruan H, Hayes-Gill B R, Mather M L *et al* 2011 Characterization of Tissue Scaffolds Using Optics and Ultrasound *Proc. SPIE* **7897** 873302
[65] Huynh N T, Parker N G, He D, Ruan H, Hayes-Gill B R, Mather M L *et al* 2011 Characterization of Tissue Scaffolds Using Optics and Ultrasound. Optical Interactions with Tissue and Cells XXII.;7897
[66] Holmes M J, Parker N G and Povey M J W 2011 Temperature dependence of bulk viscosity in water using acoustic spectroscopy *J. Phys. Conf. Ser.* **269** 012011
[67] Holmes J, Parker G, Povey W, Povey M J W, Holmes M J, Parker N G *et al* 2011 Temperature dependence of bulk viscosity in water using acoustic spectroscopy *J. Phys. Conf. Ser.* **269** 7
[68] Povey M J W, Stec Z, Hindle S A, Kennedy J D and Taylor R G 2005 Estimating organic chain length through sound velocity measurements *Ultrasonics* **43** 219–26
[69] Povey M J W, Hindle S A, Kennedy J D, Stec Z and Taylor R G 2003 The molecular basis for sound velocity in n-alkanes, 1-alcohols and dimethylsiloxanes *Phys. Chem. Chem. Phys.* **5** 73–8
[70] Jones G M, Campbell J, Povey M J W and Tong J 2001 Development of an ultrasonic oil stability monitor for the assessment of asphaltene aggregation in hydrocarbon streams *Proc. of an Int. Conf. on Mitigation of Heat Exchanger Fouling and Its Economic and Environmental Implications* pp 84–94
[71] Al-Lashi R S, Povey M J W and Watson N J Ultrasonic wave propagation in powders. 16th Anglo-French physical acoustics conference *J. Phys. Conf. Ser.* **1017** 012001
[72] Svalova A, Walshaw D, Lee C, Demyanov V, N G, Povey M J *et al* 2021 Estimating the asphaltene critical nanoaggregation concentration region using ultrasonic measurements and Bayesian inference *Sci. Rep.* **11** 1–11
[73] Gregersen S B, Povey M J W, Kidmose U, Andersen M D, Hammershoj M and Wiking L 2015 Identification of important mechanical and acoustic parameters for the sensory quality of cocoa butter alternatives *Food Res. Int.* **76** 637–44
[74] Holmes M J, Southworth T, Watson N G and Povey M J W 2014 Enzyme activity determination using ultrasound *J. Phys. Conf. Ser.* **498** 012003
[75] Parker N G and Povey M J W 2012 Ultrasonic study of the gelation of gelatin: phase diagram, hysteresis and kinetics *Food Hydrocoll.* **26** 99–107

[76] McClements D J and Povey M J W 1992 Ultrasonic analysis of edible fats and oils *Ultrasonics* **30** 383–8

[77] Hefft D I, Povey M J and Zhang Z J 2025 Transient energy released by a tribological process *Tribol. Int.* **212** 110928

Chapter 2

Wave propagation in homogeneous and inhomogeneous materials

Megan J Povey, with contributions by Valerie J Pinfield

2.1 Propagation in homogeneous materials

2.1.1 Foundations of wave propagation

To lay down the foundations for much of what follows, we first establish the essential theoretical framework for wave propagation in thermo-viscoelastic materials. We present only a summary of the main points, and refer to many classical physical acoustics texts for fuller details [1, 2]; a recent analysis of the generalised coupled wave propagation problem in thermo-viscoelastic materials can be found in Garcia Neefjes *et al* [3]. In our analysis, we consider *linear* deformations and *harmonic* oscillations, such that the stress–strain relationship follows linear elasticity (as illustrated by Hooke's law for one-dimensional strains shown in equation (2.1), generalised by use of complex, frequency-dependent moduli. In homogeneous, isotropic materials, the situation simplifies greatly, resulting in

$$\sigma^* = \lambda^* \nabla \times \xi^* + 2\mu^*\left(\frac{1}{2}\right)(\nabla\xi^* + (\nabla\xi^*)^T) \tag{2.1}$$

where the asterisk indicates a complex quantity from which a physical value is obtained in the usual way by taking the real part. ξ is the displacement, and σ is the stress (tensor). Here, we have introduced the Lamé constants generalised to complex values λ^*, μ^*; these relate to isothermal deformations. We have written the strain

$$\varepsilon = \left(\frac{1}{2}\right)(\nabla\xi^* + (\nabla\xi^*)^T) \tag{2.2}$$

For harmonic strains with time-dependence $\exp(-i\omega t)$, the rate of strain is proportional to the strain $\dot{\varepsilon} = -i\omega\varepsilon$, and, therefore, the complex moduli embody material

doi:10.1088/978-0-7503-3203-3ch2

responses typical of not only elastic solids, but also liquids or soft solids (viscous or viscoelastic materials).

The wave equations emerge from the application of conservation laws (mass, momentum, energy) using the stress–strain relationships and neglecting convection terms in the linear (low amplitude deformation) regime. Additional thermal contributions to stress can be included in addition to the elastic stresses, though these are typically negligible). The conservation equations are further analysed by use of the Helmholtz decomposition that splits the wave field into its irrotational and rotational parts, equivalent to the scalar, longitudinal waves and the vector, shear waves respectively,

$$\boldsymbol{\xi} = \nabla \varphi + \nabla \times \boldsymbol{\psi} \tag{2.3}$$

where φ is a scalar wave potential and $\boldsymbol{\psi}$ a vector wave potential. Finally, the wavenumbers $\boldsymbol{k}$ of the permitted wave propagation modes for harmonic perturbations can be determined as frequency-dependent expressions in the material properties. There exist three modes in thermo-viscoelastic media (i.e., media with non-zero thermal conductivity and with non-zero complex shear modulus). Two are longitudinal modes, irrotational in character and associated with pressure and temperature fluctuations: these are variously known as the compressional (acoustic, elastic, P-wave) mode, dominated by pressure fluctuations with relatively small temperature fluctuations and the thermal mode (dominated by temperature fluctuations with relatively small pressure fluctuations). The third mode is the shear mode (S-wave), a transverse, rotational wave mode relating to the vector potential and not associated with pressure and temperature variation but having non-zero vorticity by nature.

2.1.2 Characteristics of wavenumbers

Simplified forms of the wavenumbers are shown in table 2.1 that are valid at typical operating frequencies; for any wave mode, the magnitude of the vector wavenumber is related to the speed of propagation and the attenuation for that mode, denoted m(C, T or S to indicate compressional, thermal or shear modes), as follows:

$$k_m = \frac{w}{c_m} + i\alpha_m \tag{2.4}$$

such that the (real) speed of propagation relates to the real part of the wavenumber, $\mathrm{Re}(k_m) = w/c_m$, while the attenuation (i.e., amplitude loss) relates to the imaginary part of the wavenumber, $\mathrm{Im}(k_m) = \alpha_m$.

Note that the wave propagation is adiabatic at frequencies in water up to around 10^{14} Hz so a measured wave speed relates to the adiabatic elastic moduli (see further notes below on adiabatic and isothermal moduli). An insightful physical explanation is presented by Fletcher [4]. Table 2.2 shows the relationships between the various moduli and viscosities, and between adiabatic and isothermal properties. In liquids and gases, it is very important to distinguish these two thermodynamic conditions, and at frequencies typical in liquid characterisation, the propagation of longitudinal

Table 2.1. Wavenumbers and thermal factors for wave modes in thermo-viscoelastic media.

Wave mode	Wavenumber	Thermal factor
Longitudinal compressional wave (P-wave)	$k = \frac{w}{c} + i\alpha$ where the wave speed c is related to the elastic moduli as $c = \sqrt{(K' + \frac{4}{3}G')/\rho}$ and at 'low' frequencies, the wave propagation is adiabatic, and, therefore, the speed relates to the adiabatic modulus. The attenuation relates to dissipation mechanisms encapsulated in the shear and bulk viscosities, and thermal properties, thus (the low frequency approximation) $$\alpha = \left(\frac{\omega^2}{2\rho c^3}\right)\left[\left(\eta_B + \frac{4}{3}\eta_\mu\right) + \frac{(\gamma - 1)\tau}{C_p}\right]$$	$\Gamma_c = -\frac{1}{\beta K_T}[\rho\omega^2 - (K_T + \frac{4\mu}{3})k_c^2]$ $\approx \frac{(\gamma - 1)\omega^2}{\beta c^2}$ (References Allegra and Hawley [5]) **Note**: Allegra and Hawley [5] use an expression $\frac{c_1^2}{\gamma}$ incorporating a wave speed c_1 which they define in terms of the Lamé constants which are strictly related to isothermal deformations; however, in practice, they treat c_1 as an adiabatic wave speed, so that their expression $\frac{c_1^2}{\gamma}$ correctly equates to the isothermal bulk modulus divided by the equilibrium density such that $\frac{c_1^2}{\gamma} = \frac{K_T}{\rho_0}$.
Thermal wave	$k_T = \sqrt{\frac{\omega\rho C_p}{2\tau}}(1 + i)$	$\Gamma_c = -\frac{1}{\beta K_T}[\rho\omega^2 - (K_T + \frac{4\mu}{3})k_c^2]$ $\approx -\frac{1}{\beta\tau}$
Shear transverse wave	$k_S = \omega\sqrt{\frac{\rho}{\mu^*}}$ which leads to $k_S = \omega\sqrt{\frac{\rho}{G'}}$ for an elastic solid and $k_S = \sqrt{\frac{\omega\rho}{2\eta_\mu}}(1 + i)$ for a viscous liquid with negligible real shear modulus $G' = 0$	No temperature perturbation

Table 2.2. Relationships between various viscoelastic moduli for harmonic deformations. Subscript T, S refer to isothermal and adiabatic values respectively.

Symbol	Expression	Interpretation
K^*	$K_S^* = \gamma K_T^*$	Adiabatic and isothermal bulk moduli are related using the ratio of specific heat capacities, γ.
K_T^*	$K_T{}^* = \lambda^* + \frac{2}{3}\mu^*$	Isothermal complex bulk modulus relates to Lamé constants (isothermal by definition).
K'	$K' = \frac{1}{\kappa}$	Real part of bulk modulus relates to the inverse of compressibility.
K''	$K'' = -i\omega\eta_B$	Imaginary part of bulk modulus relates to bulk viscosity.
G^*	$G^* \equiv \mu^*$ $G_T^* \equiv G_S^*$	Complex shear modulus is equivalent to the second Lamé constant (isothermal by definition). Shear modulus is identical in isothermal and adiabatic conditions.
G'	$G' = \frac{1}{C}$	Real part of shear modulus is denoted the storage modulus, and relates to the inverse of compliance.
G''	$G'' = -i\omega\eta_\mu$	Imaginary part of shear modulus is denoted the loss modulus and relates to the shear viscosity
M^*	$M^* = K^* + \frac{4}{3}G^*$ $M_T^* = \lambda^* + 2\mu^*$	Longitudinal or compression modulus (written for either isothermal or adiabatic conditions, when the corresponding moduli are used) Isothermal modulus relates to Lamé constants
M''	$M'' = K'' + \frac{4}{3}G''$ $= -i\omega(\eta_B + \frac{4}{3}\eta_\mu)$ $= -i\omega\eta_{\text{long}}$	Imaginary part of longitudinal or compression modulus relates to longitudinal viscosity

Note: Epstein and Carhart [6] use a 'coefficient of compressional viscosity' written herein as η_C which is related to those adopted more widely as: $\eta_C = \frac{3}{2}\eta_B$.

waves is adiabatic. Note that only volume-change deformations are affected by the thermal conditions, and therefore shear moduli are unaffected.

Having presented the relationship between the complex moduli, we remind ourselves of the various types of material:

- An 'ideal' liquid has non-zero compressibility (and therefore non-zero real part of bulk modulus), zero viscosities of both types, zero real part of the shear modulus. Thus $G' = G'' = \mu' = \mu'' = \eta_\mu = K'' = \eta_B = 0$.
- A viscous liquid has non-zero compressibility (and, therefore, non-zero real part of bulk modulus), non-zero viscosities, zero real part of the shear modulus. Thus $G' = \mu' = 0$.
- An elastic solid has non-zero real part of the bulk modulus and shear modulus, and zero imaginary parts to both bulk and shear moduli, i.e., zero viscosities. $G'' = \mu'' = \eta_\mu = K'' = \eta_B = 0$
- A viscoelastic material can have non-zero real and imaginary parts to all moduli.

In liquids or soft media, the wavelength of the compressional mode is much larger than the wavelengths of both the thermal and shear modes, such that

$$\left| \frac{k_c^2}{k_t^2} \right| \ll 1 \tag{2.5}$$

We also note that the nature of the wavenumbers for the thermal and shear wave modes in liquid or soft media (see table 2.1) means that these waves are highly lossy and, therefore, dissipate in only a small number of wavelengths. We are, therefore, unable to transmit thermal or shear waves through soft materials over any significant distance (more than a few micrometres at MHz frequencies) and ultrasonic characterisation methods use the compressional wave mode (P-wave). The thermal and shear modes appear only when the compressional mode interacts with a boundary with a contrasting material, for example at the surface of a droplet, bubble or particle, and they dissipate within a short distance of the interface. This characteristic of the relative size of the wavelengths has important consequences for the propagation properties in inhomogeneous materials, and the use of ultrasound as a characterisation method, as we will explore later in the chapter.

2.1.3 Wave fields in linear wave propagation

For wave propagation in the linear (low amplitude) and harmonic regime, any perturbed variable takes the form (taking displacement for a longitudinal wave as an example)

$$\boldsymbol{\xi} = \boldsymbol{\xi}_0 \exp(i\boldsymbol{k} \times \boldsymbol{r}) \exp(-i\omega t) \tag{2.6}$$

where $\boldsymbol{k}$ is the wavenumber (vector) which denotes the direction of propagation of the wave, and $\boldsymbol{\xi}_0$ is the wave amplitude of the displacement For a longitudinal wave, the displacement is parallel to the propagation direction. All variables are expressed as a perturbation over the equilibrium value; for displacement, the medium may be at rest and the equilibrium displacement is, therefore, zero, but for pressure, for example, the perturbation is additional to the equilibrium system pressure. The temperature fluctuations associated with the compressional and thermal modes are defined in relation to the displacement potential for each mode thus

$$T = \Gamma_c \varphi_c + \Gamma_t \varphi_t \tag{2.7}$$

Expressions for the wavenumbers and the respective thermal factors are presented in table 2.1; there is no temperature fluctuation associated with the shear wave.

Note: an important thermodynamic relation

$$\gamma = 1 + \frac{K_S \beta^2 T_0}{\rho_0 C_p} \tag{2.8}$$

A systematic presentation of the wave properties in generalised linear thermo-viscoelastic media has been published by Garcia Neefjes *et al* [3] and the reader is referred to that work for fuller details, particularly the wavenumbers and temperature perturbations in equations (2.22)–(2.26). Their analysis of the fully coupled problem (elastic and thermal coupling) leads to expressions that take a rather complicated form, and it remains challenging to deduce from them an analytical form in the low frequency adiabatic propagation case. There may be slight differences compared with the simplified expressions provided by Allegra and Hawley [5] that are presented here, but further work remains to be completed on this comparison.

2.2 Wave propagation in inhomogeneous materials

A consideration of wave propagation in inhomogeneous materials generally starts with a mixture rule, representing the effect of material replacement, e.g., for an oil in water emulsion, a certain proportion of the space previously occupied by water is now replaced by oil. The classic Wood equation (2.7) [7] and the Urick equation (2.8) [8] express the speed of sound through such a mixture of fluids in terms of the volume fractions of each component. However, plenty of empirical evidence demonstrates that this simple analysis is insufficient to describe the physical phenomena occurring in many inhomogeneous materials, even in the long wavelength, low frequency regime in which the theorem is proposed to be valid. Given a typical length-scale for the characteristic inhomogeneity in the material (e.g., a particle size or droplet size), the relative size of the compressional mode wavelength defines the nature of the wave propagation. Where the wavelength is long compared with the particle size, we are in the Rayleigh limit, known as the low frequency or long wavelength regime. Short wavelength propagation compared with the particle size is in the geometrical diffraction limit; and intermediate wavelengths are the most complex propagation characteristics. Our analysis here will focus on the long wavelength or intermediate wavelength regimes which are most used for characterisation methods.

For our purposes relating to ultrasonic characterisation of liquid systems, the inhomogeneous materials we will address are those consisting of particles (or droplets or bubbles) in a homogeneous liquid, such as emulsions, bubbly liquids, suspensions of solid particles. We will focus on disordered materials, i.e., well-dispersed systems, though we address the topic of order or structure later in the chapter. The theoretical model that we will adopt for our application of ultrasonic characterisation methods is multiple scattering theory. The first edition of this book presented the fundamental equations and here we present only the principles of the method and identify some key extensions. The theory is based on a model for scattering from a single scatterer (particle, droplet, cavity, etc) and subsequently considers the combined effects of scattering from many such scatterers embedded in a homogeneous host medium. At low frequency (defined by the wavelength being much larger than the scatterer size) the wave propagation through the disordered medium of scatterers behaves as if through a homogeneous medium with properties known as the 'effective' properties; these include an effective wavenumber, which is the main outcome from the theory. In the sections below, we address the stages of the theory in turn, summarising the key points and referring to recent literature.

2.2.1 Ultrasonic scattering from a single scatterer

At the heart of multiple scattering theory is a mathematical description of the scattering from a single scatterer (a particle, droplet or bubble). In 3D systems, a spherical scatterer is the simplest and, therefore, the canonical problem. By 'scattering' what we mean is that a compressional (longitudinal) wave mode (i.e., the ultrasound wave) incident on a particle results in waves of all three of the wave modes identified in section 2.2, and these waves have a directionality pattern which is not just in the forward propagation direction. The thermal and shear wave modes in liquids are also of much shorter wavelength (typically micrometres) in liquids and, therefore, they dissipate energy in a small region around the scatterer (a process defined by the imaginary part of the wavenumber). Wave energy is, therefore, lost from the forward direction of propagation, and scattered in other directions and also absorbed in the neighbourhood of the particle.

In order to define the scattering properties of a spherical object, the theoretical framework consists of the following steps:

- The incident wave is expressed as a multipole expansion of the solutions for the wave field (i.e., solutions to the Helmholtz equation in homogeneous materials) expressed in spherical polar coordinates with the origin at the centre of the sphere. The wave field (expressed as a potential—see equation (2.3)) is thereby represented as a sum of poles from monopole (breathing mode), dipole (dumb-bell) upwards, which take the form of a product of the spherical Bessel function j_n (for incoming waves) or spherical Hankel function h_n of the second kind (for outgoing waves) and the spherical harmonics expressed in terms of the associated Legendre polynomials, thus the partial wave potential

$$\varphi_n^{p,j} = z_n\left(k_p r_j\right)P_n(\cos\theta)$$

  where z_n is either j_n or h_n, for coordinates centred on scatterer j and for wave mode p and order n; each pole is referred to as a partial wave with integer order n.
- The scattered fields for each of the three wave modes, both outside and inside the scatterer are similarly expressed as multipole expansions with, as yet, unknown amplitudes.
- Boundary equations are applied at the surface of the scatterer: these are the continuity of displacement, normal and shear stresses, temperature and heat flow.
- The set of boundary equations decouples into separate poles (partial wave orders), thus resulting in a matrix equation that can be solved for the unknown partial wave amplitudes.

The scattered partial wave amplitudes external to the scatterer for unit incident partial wave amplitude are variously known as scattering coefficients or transition factors T_n^{pq} representing the amplitude of a partial wave of order n of mode q (compressional, thermal, shear) due an incident wave of order n of mode p. Thus, an incident field potential $j_n(k_p r_j)P_n(\cos\theta)$ results in a scattered field potential for mode q of $T_n^{pq} h_n(k_q r_j)P_n(\cos\theta)$.

Classic early works on this problem are by Epstein and Carhart [6] (for liquid in liquid/gas systems), Allegra and Hawley [5] (for solid particles in liquid), and Ying and Truell [9] (for cavities in solids), amongst others. A unified presentation of the model for all material types has been attempted by a number of workers, including Challis *et al* [10] and Karlsen and Bruus [11]; unfortunately there are often some approximations or errors made in relation to material properties, and it is recommended to use the generalised thermo-viscoelastic framework to construct the problem accurately.

2.2.1.1 *Transition factors: expressions and characteristics*

A selection of analytical expressions valid at low frequency are shown in table 2.3. Since *Ultrasonic Techniques for Fluids Characterisation* [12] was published, transition factors have been developed for incident waves of types other than the compressional mode, i.e., for incident thermal and shear waves; these are complementary to those previously presented which related to the incident compressional wave mode, and outgoing modes of all types. These multi-mode transition factors

Table 2.3. The transition factors (scattering coefficients) for low orders in the low frequency limit, including wave mode conversion, for liquid suspending medium, where $|k_C^2/k_T^2| \ll 1$ and $|k_C^2/k_S^2| \ll 1$.

Wave order	Transition factors (scattering coefficients)
Monopole $n = 0$ References [13]:	$T_0^{CC} = -i(k_C a)^3\left[1 - \frac{\rho'(k'_C a)^2}{\rho(k_C a)^2}\right] - i(k_C a)\frac{\Gamma_C}{\Gamma_T}\frac{\kappa'}{\kappa}\frac{(k_T a)h'_0(k_T a)(k'_T a)j'_0(k'_T a)}{h_0(k_T a)j'_0(k'_T a)D_0}\left[1 - \frac{\kappa\Gamma_T}{\kappa'\Gamma'_T}\right]^2$ $T_0^{CT} = -\frac{\kappa'}{\kappa}\frac{\Gamma_C}{\Gamma_T}\left[1 - \frac{\rho\Gamma'_C}{\rho'\Gamma_C}\right]\frac{(k'_T a)j'_0(k'_T a)}{D_0 h_0(k_T a)j_0(k'_T a)}$ $T_0^{TC} = -(k_C a)\frac{\kappa'}{\kappa}\frac{(k'_T a)j'_0(k'_T a)}{D_0(k_T a)h_0(k_T a)j_0(k'_T a)}\left[1 - \frac{\kappa\Gamma_T}{\kappa'\Gamma'_T}\right]$ $T_0^{TT} = -\frac{j_0(k_T a)}{D_0 h_0(k_T a)}\left[\frac{(k_T a)j'_0(k_T a)}{j_0(k_T a)} - \frac{\kappa'}{\kappa}\frac{(k'_T a)j'_0(k'_T a)}{j_0(k'_T a)}\right]$ where $D_0 = \left[\frac{(k_T a)h'_0(k_T a)}{h_0(k_T a)} - \frac{\kappa'}{\kappa}\frac{(k'_T a)j'_0(k'_T a)}{j_0(k'_T a)}\right]$
Dipole $n = 1$ References [14]:	$T_1^{CC} = \frac{j_n(k_C a)}{h_n(k_C a)} \times \frac{(\hat{\rho}-1)\beta_{1hS}}{D}$ $T_1^{CS} = -\frac{3j_n(k_C a)}{h_1(k_S a)} \times \frac{(\hat{\rho}-1)}{D}$ $T_1^{SC} = \frac{2i(\hat{\rho}-1)}{h_1(k_C a)(k_S a)h_1(k_S a)D}$ $T_1^{SS} = -\frac{\left[3\beta_{1jS} + 2(\hat{\rho}-1)\delta_{1jS}\right]}{D} \cdot \frac{j_1(k_S a)}{h_1(k_S a)}$

where $\hat{\rho} = \rho'/\rho$ and
$D = [3\beta_{1hS} + 2(\hat{\rho} - 1)\delta_{1hS}]$

$$\beta_{nhS} = k_S a \frac{h'_n(k_S a)}{h_n(k_S a)} - n \cdot \beta_{njS} = k_S a \frac{j'_n(k_S a)}{j_n(k_S a)} - n$$

$$\delta_{nhS} = k_S a \frac{h'_n(k_S a)}{h_n(k_S a)} + (n+1) \cdot \delta_{njS} = k_S a \frac{j'_n(k_S a)}{j_n(k_S a)} + (n+1)$$

using $n = 1$ and where prime denotes the dispersed phase and unprimed quantities the continuous phase, except on Bessel and Hankel functions where prime represents the derivative.

Higher order $n > 1$ References [15]:

$$T_n^{CC} = \frac{j_n(k_C a)}{h_n(k_C a)} \cdot \frac{n}{(n+1)} \cdot \frac{\beta_{nhS}}{\delta_{nhS}}$$

$$T_n^{CS} = -\frac{j_n(k_c a)}{h_n(k_s a)} \cdot \frac{(2n+1)}{(n+1)} \cdot \frac{1}{\delta_{nhS}}$$

$$T_n^{SC} = -\frac{in}{h_n(k_C a) h_n(k_S a)} \cdot \frac{1}{k_S a \delta_{nhS}}$$

$$T_n^{SS} = -\frac{j_n(k_S a)}{h_n(k_S a)} \frac{\delta_{njS}}{\delta_{nhS}}$$

---

are required for the extensions to account for multi-mode multiple scattering effects which are discussed later in the chapter. A further notable development is the expression of the transition factors (and later, the effective wavenumbers) in terms of dimensionless quantities particularly the product of the wavenumbers in each material (outside and inside the scatterer) and the particle radius (e.g., $k_c a$, $k_t a$, $k_s a$) together with ratios of the physical properties ($\{\hat{p}\}$) of the outer and inner materials. As we will see later in the chapter, this approach supports a better physical understanding of the behaviour of the system, and reduces the number of variables in the expressions.

Since the first edition of this book, transition factors have been developed for incident waves of types other than the compressional mode, i.e., for incident thermal and shear waves; these are complementary to those previously presented which related to the incident compressional wave mode, and outgoing modes of all types. These multi-mode transition factors are required for the extensions to account for multi-mode multiple scattering effects which are discussed later in the chapter. A further notable development is the expression of the transition factors (and later, the effective wavenumbers) in terms of dimensionless quantities particularly the product of the wavenumbers in each material (outside and inside the scatterer) and the particle radius (e.g., $k_c a$, $k_t a$, $k_s a$) together with ratios of the physical properties of the outer and inner materials. As we will see later in the chapter, this approach supports a better physical understanding of the behaviour of the system and reduces the number of variables in the expressions.

The formulae can be expressed in many different ways, due to the recurrence relations for the Bessel and Hankel functions, so these have appeared in various forms in previous references. It is worth noting that for small arguments and at order $n = 1$, it is preferable for computational purposes to use the relationship

$$x\frac{j'_1(x)}{j_1(x)} - 1 = -x\frac{j_2(x)}{j_1(x)} \tag{2.9}$$

using the higher order function (to avoid inaccuracy with the left hand side expression). The expressions presented here are valid for the Helmholtz decomposition as defined in equation (2.3); were the opposite sign convention to be adopted for the scalar term, the signs of the mode-changing transition factors T_1^{CS}, T_1^{SC} would be the opposite to those provided here. Thermal contributions were neglected when deriving the analytical forms for the first order and higher.

2.2.1.2 Features of the transition factors

The analytical expressions presented in table 2.3 are the leading order terms in the small dimensionless variable $k_C a$ which is also known as the dimensionless wavenumber; its real part (much larger than its imaginary part) is proportional (with a factor of 2π) to the ratio of the particle radius to the wavelength of the wave. Thus, small $k_C a$ relates to the long wavelength region where the particles are much smaller than the wavelength of the transmitted wave (compressional or longitudinal or P mode). However, in liquids, $|k_C^2/k_T^2| \ll 1$ and $|k_C^2/k_S^2| \ll 1$ so the thermal and shear wavelengths are much smaller than the compressional wavelength and may be similar to the particle size or smaller or larger; no assumption is made on the size of the thermal and shear wavelengths and, therefore, $|k_T a|$, $|k_S a|$ can take any magnitude.

Thermal effects contribute solely to the monopole case, where the effects of the compressibility contrast are also observed. Since the temperature fluctuations experienced in a thermal wave mode are much larger than those in the compressional mode, we have

$$\left|\frac{\Gamma_C}{\Gamma_T}\right| \approx -(\gamma - 1)\frac{k_C^2}{k_T^2} \ll 1 \tag{2.10}$$

The two contributions to the monopole coefficient T_0^{CC} are, therefore, both of order $(k_C a)^3$. The dipole coefficient for compressional to compressional mode transitions is similarly of order $(k_C a)^3$. However, it is clear that the mode-converting terms (such as compressional to thermal) have different scaling with $(k_C a)$ as follows (where $O(x^j)$ indicates 'of order x^j').

For monopole, having only compressional and thermal mode contributions:

$$\begin{aligned} T_0^{CC} &= O((k_C a)^3), \quad T_0^{CT} = O((k_C a)^2), \\ T_0^{TC} &= O((k_C a)^1), \quad \text{and } T_0^{TT} = O((k_C a)^0) \end{aligned} \tag{2.11}$$

and for dipole and higher orders, with only compressional and shear contributions

$$\begin{aligned} T_n^{CC} &= O((k_C a)^{2n+1}), \ T_n^{CS} = O((k_C a)^n), \\ T_n^{SC} &= O((k_C a)^{n+1}), \ \text{and} \ T_n^{SS} = O((k_C a)^0) \end{aligned} \tag{2.12}$$

For computational purposes, it can be convenient to use scaled coefficients or transition factors, to avoid some of the issues relating to the higher oscillatory spherical Bessel and Hankel functions especially at large arguments (i.e., the dimensionless wavenumbers). The scalings arise quite naturally from the analytical forms, but are not required for the compressional modes, when in the long wavelength region. For completeness, the scalings are summarised below [13]

$$T_{0\text{scaled}}^{CT} = T_0^{CT} \cdot h_0(k_T a) \tag{2.13}$$

$$T_{0\text{scaled}}^{TC} = \frac{T_0^{TC}}{j_0(k_T a)} \tag{2.14}$$

$$T_{1\text{ scaled}}^{CS} = T_1^{CS} \cdot h_1(k_S a) \tag{2.15}$$

$$T_{1\text{ scaled}}^{SC} = \frac{T_1^{SC}}{j_1(k_S a)} \tag{2.16}$$

$$T_{1\text{ scaled}}^{SS} = \frac{T_1^{SS} h_1(k_S a)}{j_1(k_S a)} \tag{2.17}$$

The system of boundary equations at the surface of the scatterer enables transition factors (scattering coefficients) to be determined for any material type by adopting the generalised complex moduli presented in section 2.2, for example cavities in a solid, bubbles in a liquid, liquid droplets in a liquid, solid particles in a liquid. However, the analytical forms presented in this section (table 2.3) are specific to fluid suspending/embedding media, specifically using the approximations $|k_C^2/k_T^2| \ll 1$ and $|k_C^2/k_S^2| \ll 1$ for the continuous phase. An additional restriction is that these formulations do not account for surface tension effects, and any non-linear deformations arising from resonance, e.g., for the resonant oscillations of a bubble; the expressions in table 2.3 are the low frequency limit far below the resonant condition.

2.2.1.3 Non-spherical scatterers

It is worth noting that when the wavelength is long (much larger than the scatterer size), the shape of the scatterer does not have a major impact on the scattered fields and, therefore, the scattering properties of a sphere are generally used in this regime regardless of the expected shape of the particles or droplets. At higher frequencies, the shape becomes more significant, and generally the higher the asphericity, the greater the amplitude of the higher order transition factors; in other words, higher partial wave modes or poles are required to define the scattered wave field, whereas at low frequency for spherical scatterers, the monopole and dipole orders are often sufficient.

However, scattering from inclusions of other shapes has been addressed by a number of researchers, though not always in the full thermo-viscoelastic system; it is common to neglect thermal effects, which is valid for solids systems. A triaxial ellipsoid is a useful shape that can be used to approximate certain features such as disks, rods, needles, etc [16]. The scattering from the slightly simpler spheroidal shape has been analysed for a full thermo-viscoelastic medium by Hipp [17], which provides a useful basis for application of ultrasonics for systems of non-spherical particles in such thermo-viscoelastic media. Finally, a noteworthy classic work by Anson and Chivers [18] presented the scattering from a core–shell scatterer, which can be applied to a range of coated particles. We are not attempting here to provide a literature survey for non-spherical scatterers, and merely provide pointers to classic or early works.

2.2.2 Multiple scattering of ultrasound waves

Thus far, we have considered only the interaction of an ultrasound wave with a single scatterer, in the form of a droplet, particle or bubble (far from resonance). For measurement purposes in liquid systems, we are generally interested in materials that have many such scatterers, such as a dispersion, i.e., emulsion or suspension. We must therefore establish a mathematical model to describe the propagation of an ultrasound wave through such systems of disordered inhomogeneous media comprising many individual scatterers. For the majority of our measurement systems we will work with the coherent wave, that is, the wave detected by coherent averaging of the signals. In disordered systems of weak scatterers, the scattered field produced by the many scatterers in the system combines to produce a wave which, when averaged over the possible locations of all scatterers, takes a relatively simple form and is referred to as the coherent wave. If we use a planar incident wave, the coherent wave is planar, with an effective wavenumber characteristic of the scattering medium. Thus, the disordered material (e.g., suspension) can be treated as a homogeneous medium with a set of material properties that are denoted 'effective' properties. The role of the mathematical model is to show how to relate the effective properties to the properties of the component materials and the characteristics of the medium such as particle size, concentration, etc.

The historic progression of multiple scattering models and the theoretical framework are presented rigorously elsewhere (e.g., in *Ultrasonic Techniques for Fluids Characterisation* [12], and latterly in works such as [14]). Therefore, only an outline of the method is presented here, and we then focus on developments to the models since the previous edition, particularly the extension to multi-mode scattering and multiple species. The target of the multiple scattering models when focussing on the coherent field is to obtain the effective wavenumber for the inhomogeneous system.

In summary, the steps involved in constructing the model are as follows:

- Consider a planar, linear, or harmonic wave incident on a half-space occupied by the inhomogeneous material with a planar interface (e.g., the plane $z = 0$).
- Consider a particular configuration of scatterer locations in the half-space.

- Construct an equation for the 'exciting' field at a selected scatterer, which comprises the incident field (the one that is transmitted into the medium) as well as all the fields scattered from other scatterers (which depends on the exciting field at each of those scatterers).
- This field equation is averaged over the ensemble of all possible configurations of scatterer locations; this is called ensemble averaging and is defined by the probability distribution for the scatterer locations, e.g., how likely is there to be a scatterer in one location, given that there is a scatterer at another particular location?
- The resulting set of equations is hierarchical due to this dependence of scatterer locations on other scatterer locations. An almost universally adopted method to circumvent this challenge is to apply the Lax quasi-crystalline approximation which simplifies the probabilities and truncates the set of equations so that they can be simply expressed as a single equation in terms of the pair distribution function (related to the probability of one scatterer location relative to one other).
- The form of the exciting field is then expressed as a plane wave with an effective wavenumber, K (e.g., K_C for a compressional mode); it is this effective wavenumber that is the target of the model.
- Matching terms in the equation in either the plane wave form in the effective wavenumber $\exp(iKz)$ or the wavenumber in the continuous phase $\exp(ikz)$, results in a dispersion relation for the effective wavenumber. This dispersion relation is complicated and relates to the infinite set of transition factors for the scatterers (i.e., the various partial wave orders $n = 0, 1, 2\ldots$).
- The dispersion relation can be solved numerically for the effective wavenumber as a matrix equation by truncating the wave orders appropriately. In the long wavelength region, it is often sufficient to use only monopole and dipole terms $n = 0, 1$.
- Alternatively, the dispersion relation can be expanded as a series in the volume fraction of scatterers which is assumed to be 'small' and, therefore, the series can be truncated at second order or potentially third order when working at higher concentrations.
- Once the effective wavenumber has been obtained, the effective speed of propagation and the effective attenuation for the coherent wave can be determined.

We write the expression for the effective wavenumber as follows:

$$K_C^2 = k_C^2\Big[1 + \delta_{1\phi}\phi + \phi^2\Big(\delta_{2\phi}^{CC} + \delta_{2\phi}^{CT} + \delta_{2\phi}^{CS} + \delta_{2\phi,\mathrm{corr}}^{CC} + \delta_{2\phi,\mathrm{corr}}^{CT} + \delta_{2\phi,\mathrm{corr}}^{CS}\Big) + \ldots\Big] \quad (2.18)$$

where $\varnothing$ is the volume fraction and $n_\varnothing$ is the number density of scatterers (particles, etc), related by $n_\phi = 3\varphi/(4\pi a^3)$. The terms indicated by δ_j^{pq} are all contributions to multiple scattering, that causes the effective wavenumber K_C to be different from the wavenumber k_C in the 'host' medium, the continuous phase, or the suspending medium (e.g., water for an oil in water emulsion). Subscript j indicates the order (in concentration), whilst superscripts pq refer to the modes involved in the relevant multiple scattering events at second order in concentration. In subsequent sections, we will explore the meaning, significance and expressions of these contributions, focussing on the mode-conversion contributions that have been developed since the previous edition of the book. Full expressions are provided in the highlighted references, particularly references [13, 14] and here we present the approximated forms that are valid in the long wavelength region.

Let us first begin with the contributions in the effective wavenumber that are available from classic multiple scattering models. At first order in concentration, the effective wavenumber is equivalent to that obtained by applying the independent scatterer approximation; that is, the scattering contributions from each scatterer are summed, and scatterers do not influence each other. Epstein and Carhart [6] took this approach in their work on the scattering problem and the expression can be found in many classic works (e.g., Linton and Martin [19]), thus

$$\delta_{1\phi} = -3i\sum_{n=0}^{\infty}(2n+1)\frac{T_n^{CC}}{(k_C a)^3} \approx -\frac{3i}{(k_C a)^3}[T_0^{CC} + 3T_1^{CC}] \tag{2.19}$$

The classic multiple scattering models, notably Waterman and co-workers [20, 21], Lloyd and Berry [22], and a later reworking by Linton and Martin [19] resulted in an expression for the effective wavenumber that truly accounted for the *multiple* scattering, incorporating the effect of scattered waves from all other scatterers on any single scatterer. This leads to the second order (and higher) effects in concentration. These classic models treated only the multiple scattering of the incident mode, which for liquid systems is invariably the compressional mode (P-wave). The effects are expressed by the term $\delta_{2\phi}^{CC}$, giving only the monopole and dipole contributions relevant in the long wavelength regime

$$\delta_{2\phi}^{CC} = -\frac{27}{(k_C a)^6}[T_0^{CC}T_1^{CC} + 2(T_1^{CC})^2] \tag{2.20}$$

2.2.2.1 Multi-mode multiple scattering

As noted already, classic multiple scattering considered only a single wave mode in the model for the multiple scattering effects. However, the predominant effects of the other wave modes could still be included to some extent when applying the wavenumber expressions to particular systems, through the use of the transition factor obtained by considering all possible wave modes in the host medium (see section 2.2.1). In other words, the effective wavenumber can be calculated using the transition factors for a thermo-viscoelastic medium, thus accounting for the thermal and viscous dissipation occurring in the proximity of the particle. Substituting these transition factors into the expression for the effective wavenumber enables us to account at least partially for thermal and viscous effects, even though the multiple scattering of the thermal and shear waves has not been included. Such an approach has been successfully in use for decades. However, the development of a full, multiple scattering model that incorporates the impact of the scattered thermal and shear waves on neighbouring particles has been proceeding for a long time. A number of approaches have been taken to address the problem, including those involving core–shell models in which the system is represented by a single particle embedded in the host medium (the shell) which is then embedded in an infinite medium having the effective properties of the overall medium. Such effective medium models have been reviewed by Challis, Povey and co-workers [23], and other works [14].

The effect of thermal interactions between scatterers was addressed by Isakovich [24] and was adopted into a core–shell model developed by Hemar *et al* [25] and other workers [26]. Such thermal effects are mostly observed in emulsions, i.e., liquid scatterers in a liquid medium. Another core–shell model was developed by Hipp and co-workers [27, 28], accounting for the shear wave overlap which is the primary multi-mode mechanism in a liquid suspension of solid particles. Other workers attempted to find effective properties for the liquid continuous phase that could be substituted into the classic multiple scattering model [29], but the approach failed to correctly model the limiting cases where the thermal and shear multi-mode effects are negligible (at short thermal and shear wavelengths) within the long (compressional) wavelength region.

However, experimental data continued to expose issues with the available models, indicating that further work was required in order to fully describe the effects of multi-mode multiple scattering. Development of a rigorous multiple scattering model that incorporates multiple wave modes has been undertaken over the last 15 years primarily by researchers in France (Conoir, Luppé, and Valier–Brasier) working with Norris in the USA, building firstly a 2D model (infinitely long cylindrical scatterers) [30, 31] before progressing to the 3D problem with spherical scatterers [32] and later fully reworking the method with vector additional theorems [33]. The method follows the same steps as the classic model, but retains multiple wave modes in the analysis, which requires the inclusion of all possible modes in the exciting field at any scatterer, and the 'conversion' of any incident wave mode into a scattered field of any other wave mode (due to scattering at an individual scatterer). For example, the exciting field of the compressional wave mode at one scatterer now comprises

a) the incident wave (assumed to be compressional);
b) any scattered field of compressional wave type produced at any other scatterers due to a compressional exciting field at those scatterers; and
c) any scattered fields of compressional wave type produced at other scatterers due to any other mode of the exciting field at those scatterers.

The formulation, therefore, incorporates additional summations over all possible wave conversion events at a single scatterer. An extra complication is the analysis of the vector wave fields in different spherical polar coordinate systems centred at the different scatterers; for this problem, the use of the vector addition theorems is necessitated, which are by their nature complicated and messy. The interested reader is directed towards a recent paper by Paul Martin on the subject of vector addition theorems with reworking of the problem with reference to the previously available forms [34].

For a full presentation of the model, the reader is referred to references [33]. Here, we show only the results of analytical asymptotic analysis in the long wavelength region, i.e., $|k_C a| \ll 1$ and with the assumption that $|k_C^2/k_T^2| \ll 1$ and $|k_C^2/k_S^2| \ll 1$ which is valid in liquids up to high frequencies.

For the compressional-thermal multi-mode multiple scattering, the contributions at second order in concentration can be expressed as [13]

$$\delta_{2\phi}^{CT} = \frac{9i}{(k_C a)^3} \frac{b}{a} \cdot \frac{(k_T b) h_0'(k_T b)}{[(k_C a)^2 - (k_T a)^2]} T_0^{CT} T_0^{TC} \tag{2.21}$$

and for compressional-shear multi-mode multiple scattering [35],

$$\delta_{2\phi}^{CS} = \frac{27i}{(k_C a)^3} \frac{b}{a} \frac{(k_S b) h_0'(k_S b)}{[(k_C a)^2 - (k_S a)^2]} T_1^{CS} T_1^{SC} \tag{2.22}$$

where $b = 2a$ defines the radius of the excluded volume around one particle. Strictly, $b \geqslant 2a$ and for computational purposes may be taken as slightly larger than equality.

Having presented the key expressions for the dominant multi-mode contributions to the effective wavenumber, let us now reflect on the nature of these expressions and relate them to the physical phenomena occurring in the system. Both compressional-thermal and compressional-shear contributions involve products of the wave-converting transition factors,T_n^{pq} (e.g., T_0^{CT}) that define the amplitude of a scattered wave of mode q due to an exciting field of mode p for the partial wave order n. We might interpret this product as a description of a sequence of scattering events. Let us take an example. Firstly, the incident compressional wave interacts with one scatterer, producing waves of all modes. The scattered shear wave is incident at another scatterer (i.e., it is part of the exciting field in the region of that scatterer) and produces waves of all modes when scattered by that second scatterer. This scattered compressional wave adds to the incident compressional wave (the wave transmitted into the material). The effects of many such scattering events, averaged over all scatterer configurations results in a change to the effective phase (speed) and attenuation of the compressional wave that travels through the material.

It is also worth noting that the multi-mode contributions, $\delta_{2\phi}^{CT}$ and $\delta_{2\phi}^{CS}$ are independent of $(k_C a)$ in the long wavelength region, and depend only on the thermal or shear dimensionless wavenumbers $(k_T a)$ and $(k_S a)$ respectively. This feature becomes apparent by (a) referring to the scalings for the transition factors shown in equations (2.11), (2.12), which results in cancellation of the factors of $(k_C a)^3$ in the denominator, and (b) by applying the assumption $|k_C^2/k_T^2| \ll 1$ and $|k_C^2/k_S^2| \ll 1$ which effectively reduces the square bracket in the denominator to $[-(k_S a)^2]$. Thus, $\delta_{2\phi}^{CT}$ and $\delta_{2\phi}^{CS}$ are independent of the compressional wavenumber but dependent on the dimensionless thermal and shear wavenumbers respectively in the continuous phase respectively (as well as some ratios of material properties embedded in the transition factors). As a reminder, no assumption was made regarding the magnitude of the dimensionless wavenumbers $(k_T a)$ and $(k_S a)$ which can range from small (i.e., long thermal (shear) wavelength compared with the particle radius) to large (i.e., short thermal (shear) wavelength compared with the particle radius) within the long compressional wavelength region respectively.

2.2.2.2 Effects of correlated particle position (degrees of order)

Thus far, we have not properly considered any correlations between the positions of the scatterers but we must now do so. A completely disordered system is defined by an equal probability of any scatterer being any location, independently of the position of any other scatterers. A dilute dispersion of particles (i.e., at very low concentration, that is a low number density of particles) approximates to such a condition. However, if the concentration is increased, and assuming the particles cannot intersect each other, the particle positions start to influence each other significantly, and the system is no longer uncorrelated. The simplest hard-sphere approximation describes the volume around one particle that excludes any other particles; for spherical particles, one particle centre can be no closer than $b = 2a$ to another particle centre. The results presented so far incorporate these hard-sphere effects of excluded volume, as is apparent from the appearance of the radius of the excluded volume, b in the expressions for effective wavenumber terms. However, the positional correlations have been limited to these hard-sphere effects, and therefore the probabilities of particle locations have otherwise been assumed to be independent of the position of other particles.

The correlation between any pair of particles is defined by the number density of scatterers at any location, given that there is a scatterer located at $\mathbf{r}_1$, $(n(\mathbf{r}, \mathbf{r}_1))$, thus

$$n_{\text{corr}}(\bar{r}, n_\phi) = \frac{n(\mathbf{r}, \mathbf{r}_1)}{n_\phi} - 1 \text{ for } \bar{r} = \frac{|\mathbf{r} - \mathbf{r}_1|}{b} > 1 \tag{2.23}$$

and zero otherwise, using a non-dimensionalised distance $\bar{r}$ and the average number density for the system, n_ϕ (number of particles per unit volume). For the simplest hard-sphere model, with no further correlations, $n_{\text{corr}}(\bar{r}, n_\phi) = 0$ everywhere. However, if further effects of particle positional correlations are considered, $n_{\text{corr}}(\bar{r}, n_\phi)$ will be non-zero outside the excluded volume. For well-dispersed hard-sphere systems, the pair correlation is oscillatory with amplitudes of the oscillation that are dependent on the concentration (see figure 2.1). The well-known Percus–Yevick pair correlation is an example of a correlation function for a hard-sphere system.

Having defined the correlation in particle positions, the contributions to the effective wavenumber can now be presented, for the compressional-compressional multiple scattering and for the multi-mode compressional-shear contribution.

$$\delta^{CC}_{2\phi,\ corr} = -\frac{27i}{(k_C a)^6}(L_0^{CC} + 2L_2^{CC})T_1^{CC}\,T_1^{CC} \tag{2.24}$$

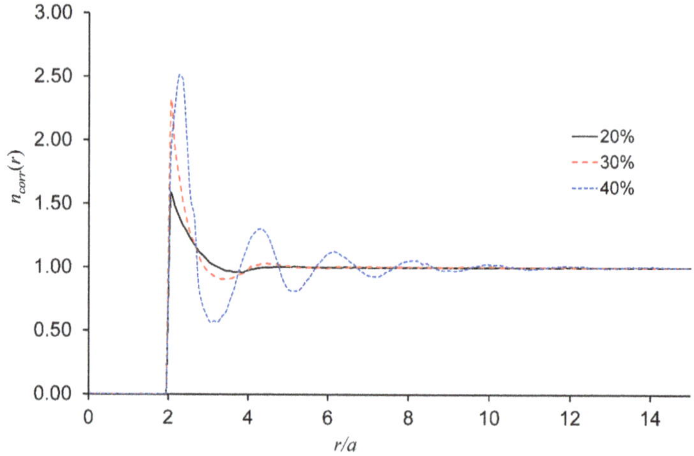

Figure 2.1. The pair correlation function for a hard-sphere system, calculated using Monte–Carlo numerical simulations of hard-sphere configurations, by Tony Valier–Brasier (CC-BY 4.0) [36].

$$\delta_{2\phi,\ corr}^{CS} = -\frac{27i}{(k_C a)^3 (k_S a)^3}\left(L_0^{CS} - L_2^{CS}\right) T_1^{CS} T_1^{SC} \tag{2.25}$$

$$L_m^{pq} = \left(k_q b\right)^3 \int_1^{\infty} j_m\left(k_p b\bar{r}\right) h_m\left(k_q b\bar{r}\right) n_{\mathrm{corr}}(\bar{r},\ n_0)\bar{r}^2 d\bar{r} \tag{2.26}$$

where the mode types p, q can be C, S relating to compressional and shear modes respectively. The correlation contribution for the multi-mode compressional-thermal contribution has not yet been separately evaluated in its asymptotic form, but the full expressions are available in [33]. The asymptotic analysis of these contributions is challenging, but the compressional term $\delta_{2\phi,\ corr}^{CC}$ is expected to be small or negligible since propagation in the long wavelength region is expected to be unaffected by the details of the configurational structure. However, asymptotic analysis demonstrates that for liquid continuous phase where the shear wave dissipates over a short distance, $\delta_{2\phi,\ corr}^{CS}$ scales only with $k_S a$ (and material property ratios in the scattering coefficients) and is independent of $(k_C a)$.

2.2.2.3 Effective wavenumber: speed and attenuation

In order to explore the effective speed and attenuation in inhomogeneous systems (which we may choose to measure to characterise our systems), we need to comprehend how the numerous variables such as frequency, particle size and material properties, affect these effective properties. We find that working in dimensionless variables is the most compact and insightful way to write our expressions and to interpret them. The relevant dimensionless variables include $(\phi,\ k_C a,\ k_T a,\ k_S a,\ \{\hat{p}\})$ where $\{\hat{p}\}$ represents the set of ratios of material properties inside and outside the particle (e.g., ρ'/ρ). These material property ratios appear within the transition factors primarily.

Combining the results presented in the previous sections with the expression for the effective wavenumber, equation (2.18), it can therefore be established that the effective wavenumber can be expressed in the form

$$K^2 = k_C^2\left[1 + G_{sq}(\phi, k_T a, k_S a, \{\hat{p}\})\right] \tag{2.27}$$

where $G_{sq}(\phi, k_T a, k_S a, \{\hat{p}\})$ is some function of the dimensionless variables in the bracket; note that this does not include $(k_C a)$. Thus, we can show that the effective speed and effective attenuation for propagation of a compressional wave through the inhomogeneous medium, in the long wavelength region, has the dependence

$$\frac{\omega}{c_{\text{eff}}} = \text{Re}(K) \approx \frac{\omega}{c_0} G_{\text{Re}}(\phi, k_T a, k_S a, \{\hat{p}\}) \tag{2.28}$$

$$(\alpha_{\text{eff}} - \alpha_0)\lambda_0 = G_{\text{Im}}(\phi, k_T a, k_S a, \{\hat{p}\}) \tag{2.29}$$

where G_{Re} and G_{Im} are two new functions. The purpose of these expressions is not to present the detail of the functions G but rather to show that the effective speed, c_{eff}, and the excess attenuation per wavelength, $\Delta\alpha \cdot \lambda_0 = (\alpha_{\text{eff}} - \alpha_0)\lambda_0$, are dependent on the volume fraction, the dimensionless wavenumbers for thermal and shear waves in the continuous (liquid) phase and the ratios of material properties (through the transition factors). α_{eff} is the effective attenuation. There is no dependence of these quantities (speed and excess attenuation per wavelength) on the dimensionless compressional wavelength, $(k_C a)$. Beyond the long wavelength region, system behaviour is decoupled and the different frequency dependence of the compressional wavenumber and the thermal/shear wavenumbers becomes important, and it therefore becomes more difficult to compare different systems, e.g., different particle sizes over a range of frequencies. In the results presented in the subsequent section, we will focus on the long wavelength region where this elegant scaling pertains.

2.2.3 Physical phenomenon of multi-mode multiple scattering

The influence of multi-mode multiple scattering effects can be interpreted by considering the thermal and shear wave decay lengths (i.e., equivalently the thermal or viscous boundary layer thicknesses) in relation to the average distance between the surfaces of particles. Multiple scattering of the thermal and shear waves will only occur to a significant extent when particles are likely to be within the thermal or shear boundary layer of another particle. The decay lengths are frequency dependent, being largest at low frequencies, and reducing to very short distances at high frequencies.

$$\delta_T = \frac{1}{\text{Im}(k_T)} = \frac{1}{\text{Re}(k_T)} = \frac{1}{\sqrt{f}}\sqrt{\frac{2\kappa}{2\pi\rho C_p}} \tag{2.30}$$

$$\delta_S = \frac{1}{\text{Im}(k_S)} = \frac{1}{\text{Re}(k_S)} = \frac{1}{\sqrt{f}}\sqrt{\frac{2\eta}{2\pi\rho}} \tag{2.31}$$

for liquids, whereas the gap between particle surfaces, $2d$, is approximately (on average)

$$2d = 2a\left(\frac{1}{\phi^{1/3}} - 1\right) \tag{2.32}$$

Figure 2.2 illustrates the relative magnitude of the viscous boundary layer (shear decay length) which is frequency dependent, in comparison with the frequency-independent inter-particle distance which is instead concentration-dependent. Any condition in which the inter-particle distance is smaller than the shear decay length has the potential to exhibit multi-mode scattering effects.

On the basis of distance alone, therefore, we might expect to see the effects of multi-mode multiple scattering at low frequencies, where the decay lengths are long, and at high concentrations when the inter-particle distances are short. However, the distances are not the only consideration. Typically, the transition factors have their highest magnitude in the region where the wavelength (of the thermal or shear wave) is similar to the particle size, and the strength of the scattered field falls either side of this resonant condition. Thus at low frequency, despite the long decay lengths that promote multiple scattering through the influence of scattered fields on neighbouring particles, the amplitude of those scattered thermal and shear wave fields is in fact small (at small $\text{Re}(k_T a)$ or $\text{Re}(k_S a)$) and therefore there is very little impact of the thermal and shear multiple scattering effects. It is only when near the resonant condition ($\text{Re}(k_T a) \approx 1$ or $\text{Re}(k_S a) \approx 1$) that the amplitude of the scattered thermal or shear waves from a single particle is sufficiently large for these fields to make a significant contribution to multiple scattering effects at neighbouring particles.

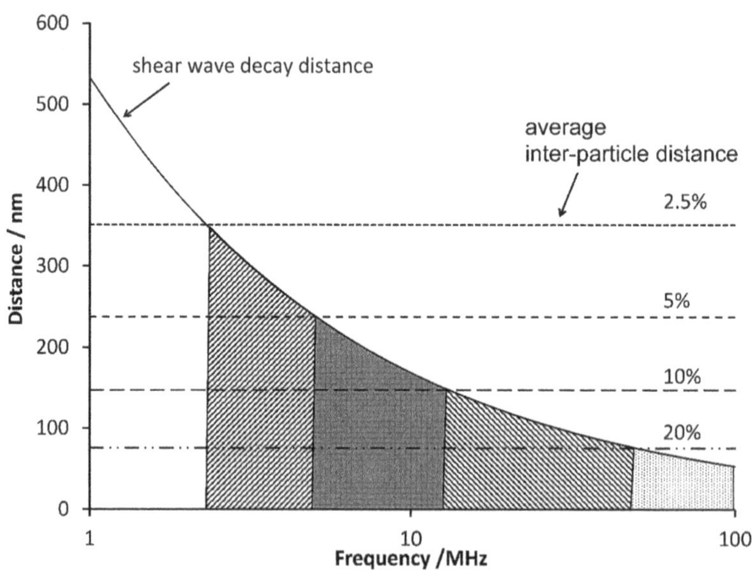

Figure 2.2. A comparison of the shear wavelength in water as a function of frequency and the inter-particle distance (between particle surfaces) at different concentrations.

The amplitude of the scattered fields is also determined by the contrast in material properties between the suspending medium and the particles (e.g., the density ratio), denoted by the set of material property ratios $\{\hat{p}\}$ in equations (2.28), (2.29).

We, therefore, deduce that the conditions under which multi-mode multiple scattering effects will influence the effective speed and attenuation are affected by the concentration ϕ (that affects the inter-particle distance), the dimensionless thermal or shear wavenumbers $(k_T a)$, $(k_S a)$ that determine the decay lengths, and the material property ratios $\{\hat{p}\}$ that determine the amplitude of the scattering. The dimensionless variables $(k_T d)$, $(k_S d)$ encapsulate the relationship of the decay length to the inter-particle distance, but these are not independent of the already-defined dimensionless variables (see equation (2.32)).

2.2.4 Effective speed and attenuation in emulsions and suspensions

Having set out in physical terms the broad dependencies of the multi-mode effects, we can now examine the effective properties in more detail. The overall behaviour of effective speed and attenuation with frequency (through $\text{Re}(k_T a)$ or $\text{Re}(k_S a)$) is similar in emulsions of liquid particles, in which thermal effects dominate, and suspensions of solid particles, in which visco-inertial effects dominate. The broad features have been described in many previous works, and here we bring out only the main points.

Firstly, we consider emulsions of liquid particles, where the multi-mode scattering is predominantly mediated by monopolar thermal fields, that dissipate in the region around each particle, but that can be scattered by neighbouring particles that are sufficiently close. Figure 2.3(a) shows the dependence of effective speed and the excess effective attenuation per wavelength as a function of the dimensionless

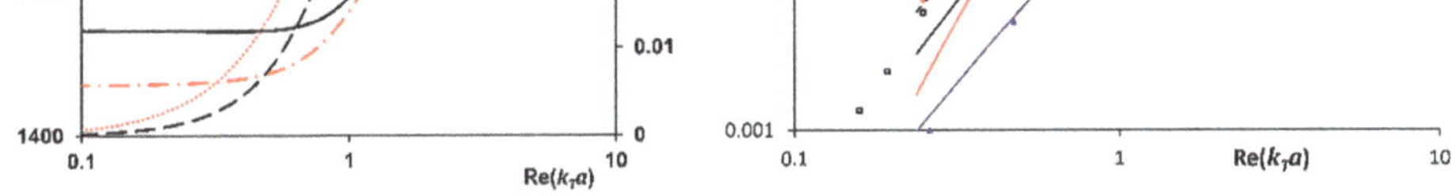

Figure 2.3. Effective speed and excess effective attenuation per wavelength (in water) against the real part of the dimensionless thermal wavenumber. (a) for a 30 v/v% bromohexadecane in water emulsion with particle radius 0.1 μm. The red lines are for classic multiple scattering theory (MST), whilst the black lines are for the multi-mode scattering model with thermal scattering effects included. The speeds are plotted as red dash-dot line (classic MST), black solid line (thermal multi-mode model). Excess attenuation per wavelength are the red dotted line (classic MST) and the black dashed line (thermal multi-mode model). The same dataset was presented in [13]; (b) for a hexadecane in water emulsion with particle radius 0.5 μm at various concentrations. The same dataset was presented in [37]. Dashed lines show classic MST, while solid lines show the thermal multi-mode model. Symbols show experimental data originally from McClements *et al* [26]. The reader is referred to the original papers for the material properties.

thermal wavenumber $\mathrm{Re}(k_T a)$, which is proportional to $a\sqrt{f}$ (see table 2.1 for the wavenumber expression). Both effective speed and excess effective attenuation per wavelength tend towards the classic multiple scattering limit at high frequency, or equivalently, high values of $\mathrm{Re}(k_T a)$. This is the region in which the thermal wavelength is small compared with particle size and the thermal dissipation around the particle is in a region very close to the particle. Therefore, there is very little influence the thermal field on neighbouring particles, and the scattered field is identical to the case where multiple thermal scattering effects are neglected (i.e., classic multiple scattering theory (MST)).

At low frequency, the effective speed tends to a constant value, but the thermal multi-mode scattering effects cause this speed to be higher (in this case) than it would be if thermal overlap effects are neglected. The low frequency region corresponds to long thermal wavelengths, so the region of thermal dissipation around each particle is rather large and can affect neighbouring particles. In this region, the attenuation reduces to zero, just as it does in classic MST, since the amplitude of the scattered thermal field itself becomes negligible as the frequency decreases (smaller $\mathrm{Re}(k_T a)$). Between these low and high limits of $\mathrm{Re}(k_T a)$ there is a resonance condition, in which the excess attenuation per wavelength has a peak, and the effective speed is highly dependent on $\mathrm{Re}(k_T a)$(i.e., on frequency for a particular system with a defined particle size).

Figure 2.3(b) shows a comparison of the attenuation for the two models, with some experimental data from other works. The experimental data generally has a lower attenuation per wavelength at the peak than predicted by the classic MST model, and the thermal multi-mode scattering model also reduces the attenuation peak but only marginally. There seems to be some improvement (i.e., agreement with experimental data) by the inclusion of thermal multiple scattering terms in the small $\mathrm{Re}(k_T a)$ (i.e., low frequency) region, but there is insufficient data to be convincing, and the model appears to exhibit some unphysical dependencies at high concentrations (see later comments).

Typically, experimental data is plotted directly as a function of frequency, rather than the dimensionless variables that best encapsulate the system behaviour; figure 2.4 illustrates such spectra for an emulsion system. We have established that the peak of the excess attenuation per wavelength is lower when multi-mode scattering is considered compared with the classic MST with no multi-mode effects (figure 2.3(a)). However the position of the peak also shifts to higher frequency MST (higher $\mathrm{Re}(k_T a)$) due to multi-mode effects (figure 2.3(a)) and the consequence of that shift is that the actual effective attenuation at a particular frequency may in fact be either higher or lower than the classic due to multi-mode effects. Figure 2.4(b) shows a spectrum of effective attenuation against frequency, demonstrating in this case that the attenuation is reduced by thermal multiple scattering effects over most of the frequency range. The speed is most affected by the multi-mode effects at the lowest frequencies (figure 2.4(a)).

For suspensions of solid particles, the dominant multi-mode effects are generally due to the shear waves, and thermal effects are negligible. The dipolar resonance that gives a peak in the excess attenuation per wavelength is governed by visco-

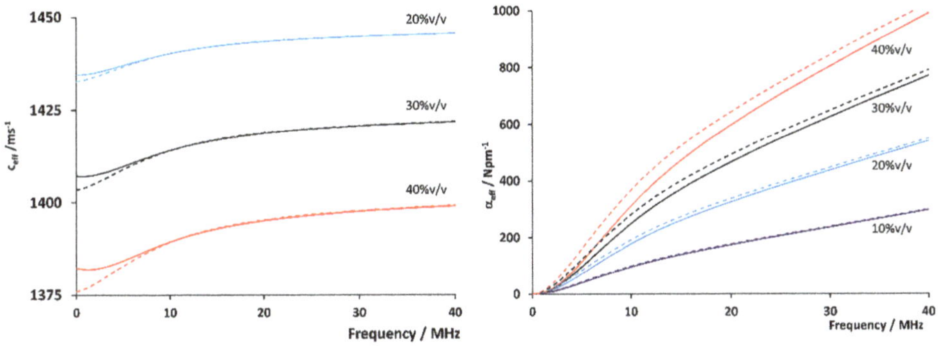

Figure 2.4. Effective speed and effective attenuation as a function of frequency at different concentrations for emulsions of bromohexadecane in water with particle radius 0.1 μm. The dataset was presented in a similar form in [37]. Dashed lines are the classic MST model, and solid lines are the multi-mode scattering model with thermal multiple scattering included.

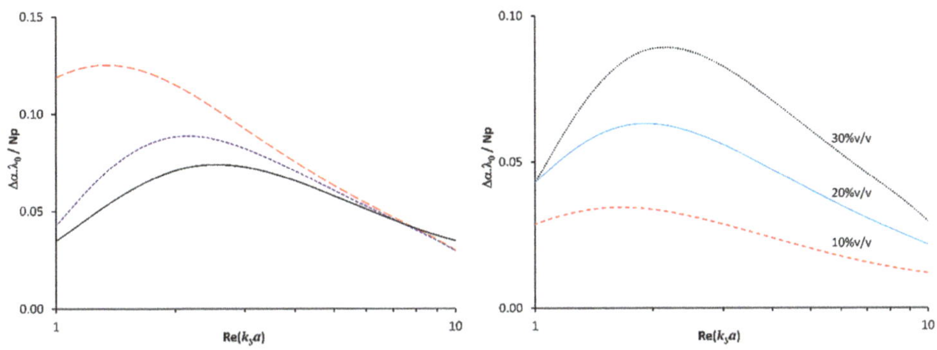

Figure 2.5. Excess effective attenuation per wavelength for silica in water suspension with particle radius 0.5 μm. (a) classic MST (red dashed line), multi-mode model (purple short-dashed line), multi-mode model with correlations (black solild line). (b) multi-mode multiple scattering with no particle correlations apart from excluded volume, showing effects of concentration; red dashed line (10%v/v), blue line (20%v/v), black solid line (30%v/v). The dataset was presented in [36] and is openly available using the DOI: 10.17028/rd.lboro.25784238.v3.

inertial effects, through the competing effects of the density contrast (inertial effects), and the viscous drag due to the surrounding liquid. Figure 2.5(a) shows the classic MST prediction, in comparison with the multi-mode model, showing a dramatic reduction in the peak attenuation per wavelength when multi-mode effects are included. Physically, the phenomenon can be thought of as a recovery of the energy that would otherwise be dissipated in the region around the particles. The shear waves interacting with neighbouring particles and scattered to produce compressional waves, means that energy is returned to the compressional mode and, therefore, there is a lower effective attenuation overall.

Figure 2.5(b) shows the effect of concentration on the multi-mode model prediction, without correlation effects, illustrating the increase in attenuation at higher concentrations, and the shift of the peak to higher frequency (higher Re($k_S a$))

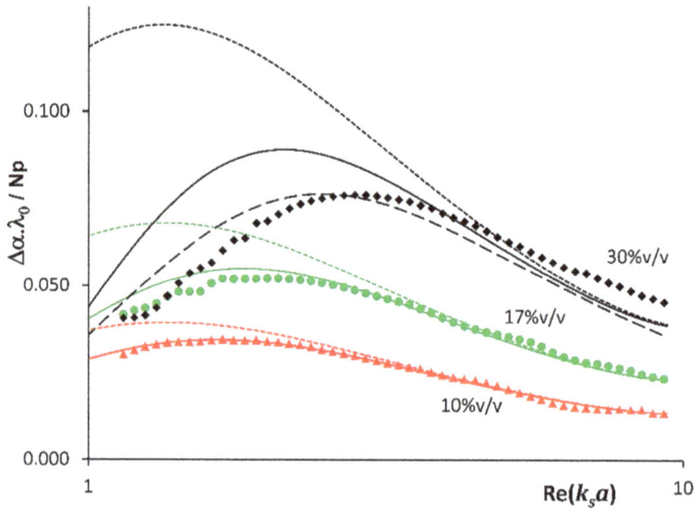

Figure 2.6. Comparison of multi-mode scattering model (solid lines) with experimental data (symbols) for excess attenuation per wavelength for silica in water suspensions with a particle radius of 0.5 μm. The solid lines are the multi-mode scattering model with no correlations (except excluded volume), the black dashed line is for the 30%v/v suspension with particle correlations, and the dotted lines are for the classic multiple scattering model with no multi-mode effects. Model presented in references [35, 36] and model comparisons with experimental data are presented in previous papers ([38, 39]). The experimental dataset is available at 10.17028/rd.lboro.2065266.v1 and 10.17028/rd.lboro.2003814.v1.

as concentration increases. Effectively, there is an increase in the effective kinematic viscosity of the suspending medium, which shifts the resonance condition, based on the effective shear wavenumber of the medium. At 30% by volume, the effect of the correlations in particle position arising from hard-sphere effects (beyond the simple excluded volume term) are quite significant (figure 2.5(a)), further reducing the peak attenuation and shifting the peak to higher frequency (higher $\mathrm{Re}(k_S a)$).

Finally, we compare the model with experimental data for silica in water suspensions (figure 2.6). It is clear that the classic multiple scattering model without multi-mode scattering effects (dotted lines) is very different from the experimental data at 30%v/v. The differences noted in the preceding paragraphs between the classic and the multi-mode multiple scattering models, are broadly confirmed by the experimental data. The data shows a much reduced peak in excess attenuation per wavelength compared with the classic multiple scattering model, and a peak position that shifts to higher $\mathrm{Re}(k_S a)$ as the concentration increases; these are the trends that the multi-mode model exhibits and are consistent with the experimental trends.

Overall, the multi-mode model is in reasonably good agreement with the experimental data up to around 20 %v/v but there are still some differences at 30%v/v. The inclusion of correlation effects (shown by the dashed black line) for the 30%v/v concentration system, improves the agreement significantly, and shifts the peak to still higher $\mathrm{Re}(k_S a)$ and reduces the peak attenuation further. The pair correlation function used in the calculations was shown in figure 2.1.

If we interpret the physical phenomena of multi-mode scattering in terms of energy recovery, we can consider the classic model to describe the complete dissipation of the energy in the scattered shear waves in the viscous boundary layer around each particle. The multi-mode model with only hard sphere excluded volume effects, describes the effect of neighbouring particles scattering the shear waves so that part of that energy is now recovered into the compressional mode, rather than being dissipated in the boundary layer. The additional correlation effects for a simple hard-sphere system (figure 2.1) are seen to reduce the peak attenuation even further, an effect that is probably due to the increased likelihood of particles being very close to each other (i.e., where the pair correlation is greater than unity at the closest separations, figure 2.1). Since the shear waves decay over a very short distance (around a couple of micrometres), it is only those particles that are very close together that will contribute to the multi-mode scattering significantly, and, therefore, it is this higher probability of very close positions that contributes to the reduction in the peak excess attenuation per wavelength.

As yet, we have not conducted an investigation of the application of the model to agglomerated systems to determine whether a further reduction in peak attenuation is observed. However, the impact of correlations on the attenuation does indicate the opportunity for attenuation spectroscopy to characterise some properties of the partial structuring in dispersed liquid systems. For example, it may be possible to deduce some properties of the structure factor or pair correlation function. Since the shear wavelength (or viscous boundary layer thickness) is frequency dependent, a spectroscopic technique effectively probes different distances in the system.

2.2.5 Multiple dispersed species

Most multiple scattering models are predicated on the assumption that all scatterers are identical. In practice, we are likely to be interested in systems where this is not case, even if that is solely due to polydispersity, i.e., different sizes of scatterer. Pragmatically, a method for accounting for polydispersity was introduced in the 1990s [40] in which the transition factor (for any order n) is replaced by a number-averaged transition factor

$$\bar{T}_n^{CC} = \sum_{j=1}^{\chi} T_{n,j}^{CC} n_j = \sum_{j=1}^{\chi} \frac{T_{n,j}^{CC} \phi_j}{a_j^3} \left(\frac{3}{4\pi} \right) \tag{2.33}$$

where the summation is over the size 'bins' (there are an integer χ bins), with particle radius a_j with a transition factor for that size of particle $T_{n,j}^{CC}$, and the volume fraction (in the whole dispersion) of particles in the size bin is ϕ_j. The model was a modification to the classic multiple scattering model in which no multi-mode effects were considered, so the only transition factors are only the compressional-compressional terms. The same approach can be, and has been, applied to systems with different dispersed phase species beyond particle size differences, e.g., an emulsion which also contains solid particles.

In more recent times, the group of Abrahams and Parnell have rigorously addressed the multiple scattering problem in systems with multi-species, extending beyond particle size to arbitrary species with a different transition factor for every species, for a system of χ different species [41] They reworked the multiple scattering model (for single mode systems, no multi-mode effects) and obtained an expression for the effective wavenumber in such systems. It is of necessity somewhat more complicated than the classic models, due to the tracking of different species with their associated transition factors, and accounting for every possible combination of species when analysing the pairs of particles. However, an analysis by Pinfield (available online as a conference presentation) [42] has demonstrated that, in the long wavelength region, the expression is equivalent to the use of the number-averaged transition factor (equation (2.33)). The group [41, 43] identified some interesting and perhaps surprising effects arising in bimodal systems with very different sizes, that are worthy of further investigation.

As yet, we are not aware of any analysis of the multiple-species system in multi-mode media. Although there are no new physical phenomena to consider, the expressions are expected to become more complicated and extensive; however, this is certainly a useful area for future work so that system of mixed dispersed species can be characterised effectively.

2.2.6 Other models of wave propagation

Although we have not attempted in this chapter to provide an extensive and comprehensive literature review of theoretical work on ultrasonic propagation in liquid systems, we now provide a few pointers to other approaches that may be of interest. In this chapter we have focussed on the coherent wave propagation in the long wavelength region, which is useful for characterisation methods and enables some simplified expressions to be presented. We have also assumed weak scattering even though weak resonances exist due to the thermal and shear effects.

However, other models exist, such as the coupled phase hydrodynamic models. It is, therefore, worth mentioning the work of Valier–Brasier Conoir and Coulouvrat [44] who demonstrated the equivalence of the results of multiple scattering models and the coupled phase models. In addition, whilst we have focussed on coherent wave propagation (leading to effective properties), the multiple scattering problem can be examined in other ways, such as through the incoherent part of the scattered field. For example, John Page's group have investigated ballistic and diffusive regimes of scattering, and latterly have worked with Thomas Brunet to examine these effects in soft materials of resonant scatterers [45]. Further details on their methods can be obtained by exploring their earlier works [46].

Resonant scattering is another topic that has grown in importance over recent years, with interesting phenomena arising from the strong scattering that occurs in such systems. Exploitation of the resonant scattering has emerged as part of the metamaterials field. Although metamaterials are largely solid systems, applications in viscoelastic media have been explored for use in sound absorbing stealth coatings. Since the nature of the interaction between scatterers is due to a viscous boundary layer in these systems, they have many characteristics in common with those observed in the multi-mode scattering in liquid suspensions of solid particles presented in section 2.2. See, e.g., recent works on the theory [47] and the implementation of these principles for material design [48].

References

[1] Pierce A D 1994 *Acoustics: An Introduction to its Physical Principles and Applications* (New York: McGraw Hill) p 678

[2] Morse P M and Ingard K U 1968 *Theoretical Acoustics* (Princeton: McGraw Hill) p 927

[3] García Neefjes E, Nigro D, Gower A L, Assier R C, Pinfield V J and Parnell W J 2022 A unified framework for linear thermo-visco-elastic wave propagation including the effects of stress-relaxation *Proc. R. Soc. A. Math., Phys. Eng. Sci.* **478** 20220193

[4] Fletcher N H 1974 Adiabatic assumption for wave propagation *Am. J. Phys.* **42** 487–9

[5] Allegra J R and Hawley S A 1972 Attenuation of sound in suspensions and emulsions: theory and experiments *J. Acoust. Soc. Am.* **51** 1545–64

[6] Epstein P S and Carhart R R 1953 The absorption of sound in suspensions and emulsions. I. Water fog in air *J. Acoust. Soc. Am.* **25** 553–65

[7] Wood A B 1964 *A Textbook of Sound* 3rd edn (London: Bell and Sons)

[8] Urick R 1947 A sound velocity method for determining the compressibility of finely divided substances *J. Appl. Phys.* **18** 983–7

[9] Ying C F and Truell R 1956 Scattering of a plane longitudinal wave by a spherical obstacle in an isotropically elastic solid *J. Appl. Phys.* **27** 1086–97

[10] Challis R E, Tebbutt J S and Holmes A K 1998 Equivalence between three scattering formulations for ultrasonic wave propagation in particulate mixtures *J. Phys. D: Appl. Phys.* **31** 3481–97

[11] Karlsen J T and Bruus H 2015 Forces acting on a small particle in an acoustical field in a thermoviscous fluid *Phys. Rev. E—Stat. Nonlin. Soft Matter Phys.* **92** 043010

[12] Povey M J W 1997 *Ultrasonic Techniques for Fluids Characterization* (San Diego, CA: Academic) p 214

[13] Pinfield V J 2014 Thermo-elastic multiple scattering in random dispersions of spherical scatterers *J. Acoust. Soc. Am.* **136** 3008–17

[14] Pinfield V J and Forrester D M 2017 Multiple scattering in random dispersions of spherical scatterers: effects of shear-acoustic interactions *J. Acoust. Soc. Am.* **141** 649–60

[15] Alam M M, Pinfield V J and Maréchal P 2020 Scattering coefficients for a sphere in a visco-acoustic medium for arbitrary partial wave order *Wave Motion* **97** 102589

[16] Dassios G 1977 Convergent low-frequency expansions for penetrable scatterers *J. Math. Phys.* **18** 126–37

[17] Hipp A K 2009 A model for sound absorption by spheroidal particles *J. Acoust. Soc. Am.* **125** 3526–38

[18] Anson L W and Chivers R C 1993 Ultrasonic scattering from spherical shells including viscous and thermal effects *J. Acoust. Soc. Am.* **93** 1687–99

[19] Linton C M and Martin P A 2006 Multiple scattering by multiple spheres: a new proof of the Lloyd–Berry formula for the effective wavenumber *SIAM J. Appl. Math.* **66** 1649–68

[20] Fikioris J G and Waterman P C 1964 Multiple scattering of waves. II. 'Hole corrections' in the scalar case *J. Math. Phys.* **5** 1413–20

[21] Waterman P C and Truell R 1961 Multiple scattering of waves *J. Math. Phys.* **2** 512–37

[22] Lloyd P and Berry M V 1967 Wave propagation through an assembly of spheres: IV. Relations between different multiple scattering theories *Proc. Phys. Soc.* **91** 678

[23] Challis R E, Povey M J W, Mather M L and Holmes A K 2005 Ultrasound techniques for characterizing colloidal dispersions *Rep. Prog. Phys.* **68** 1541–637

[24] Isakovich O, Rasprostranenii Zvuka V and Emulsiyakh 1948 Hurnal Eksperimentalnoi I Teoreticheskoi Fiziki**18** 907–12 18:907-12

[25] Hemar Y, Herrmann N, Lemaréchal P, Hocquart R and Lequeux F 1997 Effective medium model for ultrasonic attenuation due to the thermo-elastic effect in concentrated emulsions *J. Phys. II* **7** 637–47

[26] McClements D J, Hemar Y and Herrmann N 1999 Incorporation of thermal overlap effects into multiple scattering theory *J. Acoust. Soc. Am.* **105** 915–8

[27] Hipp A K, Storti G and Morbidelli M 2002 Acoustic characterization of concentrated suspensions and emulsions. 1. Model analysis *Langmuir* **18** 391–404

[28] Hipp A K, Storti G and Morbidelli M 2002 Acoustic characterization of concentrated suspensions and emulsions. 2. Experimental validation *Langmuir* **18** 405–12

[29] Challis R E and Pinfield V J 2014 Ultrasonic wave propagation in concentrated slurries—the modelling problem *Ultrasonics* **54** 1737–44

[30] Conoir J-M and Norris A N 2010 Effective wavenumbers and reflection coefficients for an elastic medium containing random configurations of cylindrical scatterers *Wave Motion.* **47** 183–97

[31] Norris A N and Conoir J-M 2011 Multiple scattering by cylinders immersed in fluid: high order approximations for the effective wavenumbers *J. Acoust. Soc. Am.* **129** 104–13

[32] Luppé F, Conoir J-M and Norris A N 2012 Effective wave numbers for thermo-viscoelastic media containing random configurations of spherical scatterers *J. Acoust. Soc. Am.* **131** 1113–20

[33] Luppé F, Conoir J-M and Valier-Brasier T 2022 Longitudinal and transverse coherent waves in media containing randomly distributed spheres *Wave Motion.* **115** 103082

[34] Martin P A 2024 Another look at addition theorems for vector spherical wavefunctions *Math. Methods Appl. Sci.* **47** 12443–59

[35] Pinfield V J and Valier-Brasier T 2024 Multi-mode multiple wave scattering in suspensions of solid particles in viscous liquids: part 1 asymptotic results *Proc. R. Soc. A. Math., Phys. Eng. Sci.* **480** 20240083

[36] Pinfield V J and Valier-Brasier T 2024 Multimode multiple wave scattering in suspensions of solid particles in viscous liquids: part 2: numerical validation *Proc. R. Soc. A. Math., Phys. Eng. Sci.* **480** 20240106

[37] Pinfield V J 2014 Advances in ultrasonic monitoring of oil-in-water emulsions *Food Hydrocoll.* **42** 48–55

[38] Forrester D M, Huang J, Pinfield V J and Luppé F 2016 Experimental verification of nanofluid shear-wave reconversion in ultrasonic fields *Nanoscale* **8** 5497–506

[39] Forrester D M, Huang J and Pinfield V J 2016 Characterisation of colloidal dispersions using ultrasound spectroscopy and multiple-scattering theory inclusive of shear-wave effects *Chem. Eng. Res. Des.* **114** 69–78

[40] McClements D J 1991 Ultrasonic characterisation of emulsions and suspensions *Adv. Colloid Interface Sci.* **37** 33–72

[41] Gower A L, Smith M J A, Parnell W J and Abrahams I D 2018 Reflection from a multi-species material and its transmitted effective wavenumber *Proc. R. Soc. A. Math., Phys. Eng. Sci.* **474** 20170864

[42] Pinfield V (ed) 2023 Multiple wave scattering in soft complex media *British Applied Mathematics Colloquium* (Loughborough University)

[43] Gower A L, David Abrahams I and Parnell W J 2019 A proof that multiple waves propagate in ensemble-averaged particulate materials *Proc. R. Soc. A. Math., Phys. Eng. Sci.* **475** 20190344

[44] Valier-Brasier T, Conoir J-M, Coulouvrat F and Thomas J-L 2015 Sound propagation in dilute suspensions of spheres: analytical comparison between coupled phase model and multiple scattering theory *J. Acoust. Soc. Am.* **138** 2598–612

[45] Tallon B, Brunet T and Page J H 2023 Ultrasonic wave transport in concentrated disordered resonant emulsions *Phys. Rev.* B **108** L060202

[46] Leroy V, Strybulevych A, Page J H and Scanlon M G 2011 Influence of positional correlations on the propagation of waves in a complex medium with polydisperse resonant scatterers *Phys. Rev. E.* **83** 046605

[47] Simon A, Baudis Q, Wunenburger R and Valier-Brasier T 2024 Propagation of elastic waves in correlated dispersions of resonant scatterers *J. Acoust. Soc. Am.* **155** 3627–38

[48] Sharma G S, Skvortsov A, MacGillivray I and Kessissoglou N 2023 Sound absorption by a soft medium embedded with complex-shaped hard inclusions *J. Acoust. Soc. Am.* **154** A124–A

Chapter 3

Rheology, acoustics, elasto-dynamics, and the complex moduli

Megan J Povey, with contributions by Daniel I Hefft

Visco-elastic fluid materials, represent an intriguing class of substances as their mechanical properties lie between those of traditional rigid solids and viscous fluids. Everyday examples include melting chocolate, yoghurt, toothpaste, and tomato ketchup. These materials exhibit both solid-like characteristics, such as maintaining their overall shape, and liquid-like behaviour, allowing for significant deformations under applied stress [1]. The deformation behaviour of soft solids is typically quantified by the stress–strain relationship, which describes how the material responds to external mechanical forces. In the case of linearly elastic materials, Hooke's law relates 1D stress, σ, and 1D strain, ε, through a constant called Young's modulus, E:

$$\sigma = E \cdot \varepsilon \tag{3.1}$$

Young's modulus signifies the material's stiffness and how resistant it is to deformation. However, for soft solids, this relationship can become non-linear at larger strains due to the rearrangement and interactions of their microstructures. Non-linear models, such as the Ogden or neo-Hookean models, are employed to describe their stress–strain behaviour more accurately. The Ogden model [2] is based on the strain energy density function, W, which describes the energy stored in a material due to deformation. The Ogden model can accommodate the complex stress–strain behaviour observed in soft solids, making it suitable for a wide range of applications in biomedical engineering, soft robotics, and material design. However, for practical reasons, the neo-Hookean model is often used as it is a simpler representation of the non-linear behaviour of soft solids [3] and is particularly useful for materials that exhibit nearly incompressible behaviour. The neo-Hookean model provides a good approximation for small to moderate deformations in soft solids

doi:10.1088/978-0-7503-3203-3ch3

and has been widely used in engineering towards certain microstructure properties. One of the distinctive characteristics of soft solids is their visco-elastic behaviour, which combines features of viscosity and elasticity. When subjected to stress, soft solids exhibit time-dependent responses. The relaxation modulus, $G(t)$, is a fundamental parameter in characterising visco-elasticity. It quantifies the material's ability to relax and dissipate stress over time, indicating the gradual return to its original state after deformation. Mathematically, the relaxation modulus is related to the stress relaxation function, $R(t)$, representing the stress decay with time, through the transform:

$$G(t) = \int R(t)dt \tag{3.2}$$

The stress relaxation function relies on the material's internal dynamics, which are often influenced by the arrangement and interactions of its constituents, such as polymer chains, colloidal particles, or biological fibres. Soft solids also exhibit creep behaviour, which refers to their time-dependent deformation under a constant applied stress. The creep compliance, $J(t)$, characterises the material's strain response over time when subjected to constant stress. Like the relaxation modulus, the creep compliance is related to the creep compliance function, $C(t)$, representing the strain development with time under constant stress, through the transform:

$$J(t) = \int C(t)dt \tag{3.3}$$

Creep can have significant implications in engineering applications where materials must withstand prolonged loads, as in building foundations or visco-elastic damping materials. Rheology is a crucial aspect of soft solids, dealing with their flow and deformation behaviour. Soft solids' visco-elastic nature leads to complex flow responses, particularly under oscillatory stress. The complex viscosity, η^*, is a key parameter in rheology, representing the material's resistance to flow [4].

The complex viscosity comprises two components: the storage modulus, η', characterising the elastic response of the material, and the loss modulus, η', representing the viscous response

$$\eta^* = \eta' + i\eta'' \tag{3.4}$$

The storage modulus represents the material's ability to store and recover elastic energy, while the loss modulus accounts for the dissipation of energy as heat during deformation.

Ultrasound waves are mechanical waves that propagate through a medium, including soft matter foods, by transferring energy through particle oscillations. These waves exhibit longitudinal compressional behaviour, where particles vibrate parallel to the direction of wave propagation. In the case of soft matter foods, these particles represent the constituents of the material, such as water, oil, or protein molecules.

The behaviour of mechanical waves, including ultrasound, is governed by the wave equation, which describes how the wave's amplitude evolves over both time and space. For one-dimensional wave propagation, the wave equation can be expressed as:

$$\frac{\partial^2 \xi(x, t)}{\partial t^2} - \frac{c^2 \partial^2 \xi(x, t)}{\partial x^2} = 0 \tag{3.5}$$

In this equation, $\xi(x, t)$ represents the displacement of particles at position x and time t. The first term, $\frac{\partial^2 \xi}{\partial t^2}$, represents the acceleration of particles over time, and the second term, $\frac{c^2 \partial^2 \xi}{\partial x^2}$, represents the curvature of the wavefront in space. The constant c is the speed of sound in the medium, which depends on the material's properties, such as density and compressibility. For soft matter foods, which are typically visco-elastic, the wave equation can be modified to incorporate the effects of visco-elasticity. The generalised wave equation becomes:

$$\frac{\partial^2 \xi}{\partial t^2} - \frac{c^2 \partial^2 \xi}{\partial x^2} = \eta \frac{\partial \xi}{\partial t} + \frac{\eta' \partial^2 \xi}{\partial t^2} \tag{3.6}$$

In this modified equation, η and η' represent the damping coefficients that characterise the visco-elastic behaviour of the material. These coefficients account for the energy dissipation and the delayed response of the material to mechanical perturbations. When ultrasound waves propagate through a soft matter (food), they interact with its microstructure, leading to changes in wave behaviour. As the ultrasound waves encounter interfaces between different phases within the food, such as solid–gel or liquid–air interfaces, reflections and refractions occur due to acoustic impedance mismatches. This acoustic impedance, Z, at an interface between two media is defined as the product of the material's density, ρ, and the speed of sound, c:

$$Z = \rho \times c \tag{3.7}$$

The variation in acoustic impedance at interfaces causes echoes and can be utilised for ultrasound imaging techniques like tomography, allowing the visualisation of the soft matter food's internal microstructure. In addition to impedance mismatches, ultrasound waves experience attenuation as they traverse through the soft matter food. This attenuation is due to various phenomena, including scattering, absorption, and visco-elasticity. Scattering occurs when the waves encounter inhomogeneities or microstructural elements within the soft matter food, causing changes in the direction of wave propagation. Absorption involves the conversion of ultrasound energy into heat as it interacts with the constituents of the material. Visco-elastic properties of soft matter foods also contribute to attenuation, with energy being dissipated as the waves interact with the material's structure. For a detailed description of ultrasound scattering see Povey [5]. During propagation, the ultrasound wave interacts with the constituents of the soft matter, such as water molecules, proteins, and other molecules. This interaction leads to the dissipation of ultrasound energy. The rate of absorption of ultrasound energy can be quantified using the absorption coefficient, α, which represents the fraction of ultrasound energy absorbed per unit distance travelled in the soft matter.

The change in acoustic pressure, ΔP, due to absorption as the ultrasound wave travels through a small distance Δx in the soft matter can be described as:

$$\Delta P = -\alpha \times P_{\text{incident}}(x, t) \times \Delta x \tag{3.8}$$

where $\alpha \times P_{\text{incident}}(x, t)$ represents the energy absorbed per unit distance travelled and the absorption coefficient (α). The attenuation of ultrasound waves can be modelled by the attenuation equation:

$$\frac{\partial P}{\partial x} + \alpha \times P = 0 \tag{3.9}$$

In this equation, $\frac{\partial P}{\partial x}$ represents the spatial gradient of the acoustic pressure, and α is the attenuation coefficient, quantifying the rate of ultrasound intensity reduction as it propagates through the soft matter food. The detailed nature of the ultrasound attenuation coefficient is addressed in the next section.

Rheology and acoustics share a common basis in the physics and mathematics of elastic materials, and both begin from generalisations of force/stress and displacement/strain. The most general statements involve relations between a stress (force per unit area) tensor and a strain (or strain-rate) tensor, which need not be linear, isotropic or even time reversible [5, 6]. It might be thought that compressional waves involve only normal stresses whilst rheology is often about shear stresses. However normal stresses contain a term depending on viscosity and as observed by Stokes [7, 8]. the compressional wave is, therefore, linked with viscosity, mostly in the dissipation (attenuation terms). The relationship between shear viscosity η [Pa s] and longitudinal sound attenuation α_{long} [Np m^{-1}] was first derived by Stokes. Pierce [9] and Morse and Ingard [10] present a more general expression for the longitudinal attenuation α_{long} that incorporates thermal effects:

$$\alpha_{\text{long}} = \frac{\eta\omega^2}{2\rho c_{\text{long}}{}^3}\left(\frac{4}{3} + \frac{\eta_K}{\eta} + \frac{(\gamma - 1)\kappa}{\eta C_p}\right) \tag{3.10}$$

and the longitudinal viscosity is related to the shear and bulk viscosity through

$$\eta_{\text{long}} = \eta_K + 4/3\eta \tag{3.11}$$

where γ is the ratio of specific heats, κ [W m^{-1}.K^{-1}] is the thermal conductivity; C_p [J K^{-1}] is the specific heat at constant pressure; ω is radial frequency [rad s^{-1}] and η_{long} [Pa s] is longitudinal viscosity and η_K is the bulk viscosity.

Equation (3.11) is analogous to the relationship between the elastic modulus (M [Pa]), bulk modulus (K [Pa]) and shear modulus (G [Pa]) in solids

$$M = K + \frac{4}{3}G \tag{3.12}$$

Note here once again the importance of the bulk viscosity (also called the longitudinal viscosity [11]) which is measured either using ultrasound spectroscopy [12] or Brillouin scattering [11]. Equation (3.10) does not apply at extremely high frequency. The equation is low frequency because of the neglect of second order terms in the thermal and viscous contributions; the frequency range of validity is, therefore, below around 10^{14} Hz in water. It should be that other phenomena (scattering, molecular relaxation, diffraction) may form a significant component of the measured attenuation, in addition to viscosity. α_{long} is often called the classical attenuation $\propto_{classical}$. The total attenuation $\propto$ is measured by determining the reduction of signal amplitude as a function of the distance travelled by the wave x [13]. Pressure p is proportional to amplitude A or displacement ξ squared, $p \sim \xi^2$. We can then define the attenuation in terms of measurable pressure differences as

$$\alpha_{measured} = \frac{1}{x}\log_e\frac{\xi}{\xi_0}\ [\mathrm{Np\ m^{-1}}]$$

or

$$\alpha_{measured} = \frac{1}{x}20\log_{10}\frac{\xi}{\xi_0}\ [\mathrm{dB\ m^{-1}}] \tag{3.13}$$

The decibel (dB) and neper (Np) are dimensionless because they are ratios. Note that the attenuation is normally expressed in terms of pressure rather than displacement and in this case

$$\alpha\ [\mathrm{dB\ m^{-1}}] = \frac{1}{x}10\log_{10}\frac{p}{p_0} = \frac{1}{x}20\log_{10}\frac{\xi}{\xi_0} \tag{3.14}$$

In practice the relationship between the measured attenuation $\alpha_{measured}$, the classical $\alpha_{classical} = \alpha_{long}$ attenuation is far from straightforward. In particular diffraction effects (the $\frac{c^2\partial^2 u}{\partial x^2}$ term in equation (3.6)) must be accounted for [13]. The relationship between acoustic wavelength and the size of the inhomogeneities generating scattering $d_{inhomogeneity}$ is very frequency dependent and the various limits $\lambda > d_{inhomogeneity}$, $\lambda \sim d_{inhomogeneity}$ and $\lambda < d_{inhomogeneity}$ give rise to different phenomena, particularly scattering phenomena. In general,

$$\alpha_{measured} = \alpha_{long} + \alpha_{scattering} \tag{3.15}$$

Relaxation effects are incorporated into α_{long} and here note that their effects and dependence on frequency varies widely [14]. In addition, more than one relaxation process may be present. These relaxation processes complicate the comparison of ultrasonically determined viscosity and low frequency rheological determinations. In the case of a single relaxation process [5] the following expression for attenuation may be used

$$\alpha_r = \frac{C_r}{2c_{long}}\frac{\omega^2/\omega_r^2}{1+\omega^2/\omega_r^2}$$

where

$$\omega_r = \frac{2\pi}{\tau_r} \tag{3.16}$$

and α_r is the attenuation solely due to relaxation; ω is the radial frequency at which the measurement is taken, C_r is a coupling constant, c_{long} is the velocity of sound and τ_r is the relaxation time. This expression is consistent with the Maxwell model for visco-elastic behaviour [14]. *It is therefore important, in the first place, to verify the frequency-squared dependence of the attenuation and account for those mechanisms not associated with viscosity.*

In most liquids, the ratio of specific heats is close to one (in water at 100 KPa and 25 °C it is 1.0106) and the thermal term is usually neglected, so that rearranging equation (3.10) we get:

$$\eta_{\text{long}} = \frac{\alpha_{\text{long}} \rho c_{\text{long}}^3}{2\omega^2} \tag{3.17}$$

where η_{long} is defined by equation (3.11).

In rheology the moduli equations (3.12) are each expressed as two numbers, a 'real' part, relating to strains that are in step with the applied stress and an 'imaginary' part where the strains are $\pi/2$ radians out of step:

$$M = M' + iM'' = K' + iK'' + \frac{4}{3}(G' + iG'') \tag{3.18}$$

where $i = \sqrt{-1}$.

For a harmonic oscillation of the form $e^{-i\omega t}$ the strain rate is equal to the strain multiplied by a factor $-i\omega$, and the relation between complex elastic moduli and viscosity enables us to re-express the acoustic properties. In solids, the equivalent of the Newton–Laplace equation.

$$c_{\text{long}} = \sqrt{\frac{M}{\rho}} \tag{3.19}$$

In fluids and visco-elastic materials $K' \gg G'$, for example, in a yoghurt $G' \sim$ a few Pa whilst K' is $\sim 10^{10}$ Pa so that the velocity of compressional sound in liquid media is:

$$c_{\text{long}} = \sqrt{\frac{K'}{\rho}} \tag{3.20}$$

We can now rewrite equation (3.17) in terms of the complex elastic moduli:

$$\alpha_{\text{long}} = \frac{2M''}{3\rho c_{\text{long}}^3}\omega = 2\frac{K'' + 4/3G''}{3\rho c_{\text{long}}^3}\omega = \frac{2\omega^2\,\eta_{\text{long}}}{3\rho c_{\text{long}}^3} \tag{3.21}$$

$$M'' = K'' + 4/3G'' = \eta_{\text{long}}\omega \tag{3.22}$$

The above equations show that the real part of the shear modulus can contribute to the speed of compressional sound and the imaginary part of the shear modulus (which relates to the viscosity) contributes to compressional sound attenuation, clearly linking acoustics and rheology.

The relationship between the length of the longitudinal wave, and the size and geometry of the sample is important, due to additional restoring forces arising at boundaries when the longitudinal acoustic wavelength is of the order of or greater than any sample dimension. These additional forces result in acoustic parameters becoming a function of sample geometry, heterogeneity and dimensions. In the following we assume that the longitudinal wavelength is always smaller than any dimension of a sample holder.

In the field of shear acoustics [15] a highly attenuated shear wave propagates over just a few wavelengths and much more slowly [16] than the longitudinal wave. Oscillatory shear rheology [17] shares much in common with shear acoustics. The speed of an acoustic shear wave is given by [5]:

$$c_{\text{shear}}^2 = \frac{G'}{\rho} \tag{3.23}$$

and its penetration depth δ_{shear} [m]

$$\delta_{\text{shear}} = \sqrt{\frac{2\eta_{\text{shear}}}{\omega\rho}} \tag{3.24}$$

In a typical yoghurt, $c_{\text{shear}} \cong 0.03\ \text{m s}^{-1}$ whereas $c_{\text{long}} \cong 1450\ \text{m s}^{-1}$. At 1 MHz the equivalent wavelengths are $\lambda_{\text{shear}} \cong 0.03\ \mu\text{m}$ and $\lambda_{\text{long}} \cong 1.45$ mm, at 100 Hz are $\lambda_{\text{shear}} \cong 0.3$ mm and $\lambda_{\text{long}} \cong 14.5$ m. The shear penetration depth of a yoghurt with shear viscosity 2 mPa s is 2.5 mm at 1 Hz, 0.25 mm at 100 Hz, 0.025 mm at 10 kHz and 2.5 μm at 1 MHz. Schroyen [18] comments that 'The penetration depth is of particular importance at very high frequencies since it defines the maximal distance over which the sample can be probed…: At large ultrasonic frequencies (>1 MHz), the penetration depth becomes so small that surface properties rather than bulk properties are probed'.

Relationships and variables widely used in practical rheology include a complex shear modulus:

$$G^* = G' + iG'' \tag{3.25}$$

which can be expressed as (noting that G^* is formally the same as G):

$$G = |G^*|e^{i\delta} \tag{3.26}$$

with magnitude and phase given by:

$$|G^*| = \sqrt{G'^2 + G''^2} \tag{3.27}$$

$$\tan\delta = \frac{G''}{G'} \tag{3.28}$$

where δ is the phase angle between stress and strain which varies between $\pi/2$ for fluids when $G' = 0$, to 0 in the case of elastic solids where $G'' = 0$. The stress–strain relations can be written

$$G = \frac{\tau}{\gamma} \tag{3.29}$$

or

$$\eta_s = \frac{\tau}{\dot{\gamma}} \tag{3.30}$$

where τ is shear stress [Pa], γ is shear strain and $\dot{\gamma}$ is shear strain rate [s^{-1}]. In general, and particularly in soft solid materials τ and γ are complex quantities as is G.

For a harmonic strain in an oscillatory system

$$\gamma = \gamma_0 e^{i(\omega t + \delta)} \tag{3.31}$$

then

$$\dot{\gamma} = i\omega\, \gamma_0 \tag{3.32}$$

so that from equation (3.30)

$$\tau = i\omega\, \eta_s \gamma \tag{3.33}$$

therefore

$$G = i\omega\, \eta_s \tag{3.34}$$

so that in materials with no elastic component ($G' = 0$):

$$G'' = \omega \eta_s \tag{3.35}$$

References

[1] Ungar E E and Kerwin E M Jr. 2005 Loss factors of viscoelastic systems in terms of energy concepts *J. Acoust. Soc. Am.* **34** 741

[2] Yeoh O H 1997 On the ogden strain-energy function *Rubber Chem. Technol.* **70** 175–82

[3] Zhang T 2019 Deriving a lattice model for neo-Hookean solids from finite element methods *Extreme Mech. Lett.* **26** 40–5

[4] Attenburrow G, Barnes D J, Davies A P and Ingman S J 1990 Rheological properties of wheat gluten *J. Cereal Sci.* **12** 1–14

[5] Povey M J W 1997 *Ultrasonic Techniques for Fluids Characterization* (San Diego, CA: Academic) p 214

[6] Charlier J P and Crowet F 1986 Wave equations in linear viscoelastic materials *J. Acoust. Soc. Am.* **79** 895–900

[7] Stokes G G 1845 On the theories of the internal friction of fluids in motion and of the equilibrium and motion of elastic solids *Trans. Cambridge Philos. Soc.* **8** 287

[8] Stokes G G 1848 The effect of internal friction *Philos. Mag.* **33** 349

[9] Pierce A D 1994 *Acoustics: An Introduction to its Physical Principles and Applications* (New York: McGraw-Hill) p 678

[10] Morse P M and Ingard K U 1968 *Theoretical Acoustics* (Princeton, NJ: McGraw-Hill) p 927

[11] Dukhin A S and Goetz P J 2009 Bulk viscosity and compressibility measurement using acoustic spectroscopy *J. Chem. Phys.* **130** 1245191–13

[12] Ghosh S, Holmes M and Povey M 2017 Temperature dependence of bulk viscosity in edible oils using acoustic spectroscopy *J. Food Process.Technol.* **08** 1000676

[13] Povey M J W 2013 Ultrasound particle sizing: a review *Particuology* **11** 135–47

[14] Barlow A J and Lamb J 1967 Viscoelastic relaxation in supercooled liquids *Discuss. Faraday Soc.* **43** 223–30

[15] Manfredi O F, Mills R S, Schirru M M and Dwyer-Joyce R S 2019 Non-invasive measurement of lubricating oil viscosity using an ultrasonic continuously repeated chirp shear wave *Ultrasonics* **94** 332–9

[16] Dukhin A 2021 Rheology in longitudinal (ultrasound) mode *Rev. Colloid J.* **83** 1–19

[17] Weitz D, Wyss H and Larsen R 2007 Oscillatory rheology: measuring the viscoelastic behaviour of soft materials *GIT Lab. J. Eur.* **11** 68–70

[18] Schroyen B, Vlassopoulos D, Van Puyvelde P and Vermant J 2020 Bulk rheometry at high frequencies: a review of experimental approaches *Rheol. Acta.* **59** 1–22

Chapter 4

Practical measurement and analysis in ultrasound characterisation of processes

Megan J Povey, with contributions by Alexander L Bowler and Nicholas Watson

4.1 Introduction

The manufacturing sector is increasingly using real-time data collection and analysis to support decision making, improve productivity, and ensure product quality [1]. In the future, it is anticipated that supply chains, processes, and markets will eventually be fully interconnected enabling automatic optimisation in real time [2]. Sensors will play a critical role in this transformation by performing the automatic data acquisition needed to inform the decision making [3, 4]. Sensing technologies are more widely deployed in discrete manufacturing, partly due to discrete processes being simpler to monitor [5]. On the other hand, process manufacturing must measure a broader range of materials and parameters in more complex flows. For successful automated monitoring of process manufacturing systems, a wide range of sensing techniques are required. Low power (below 1 W cm^{-2}), high-frequency (above 100 kHz) ultrasonic (US) sensors provide several advantages which will make them the optimal choice for specific applications. These advantages include relatively small size, low cost, ability to monitor opaque materials, low power consumption, and real-time, non-destructive measurement capabilities. Ultrasonic sensors can also be implemented in either non-invasive or minimally invasive configurations and do not alter the structure of materials they pass through [6].

The three most common ultrasonic parameters of interest are velocity, attenuation, and acoustic impedance [6]. Ultrasonic velocity is determined by measuring the time-of-flight (TOF) of a sound wave through a known distance. Attenuation is measured by the loss of sound energy as the wave propagates through a medium. This loss can be caused by absorption in homogeneous materials (influenced by viscosity or relaxation effects) or by scattering from discontinuities in heterogeneous systems [7]. Acoustic impedance is the product of the material density and the speed

doi:10.1088/978-0-7503-3203-3ch4

of sound within that material and is typically obtained by measuring the proportion of the wave reflected at a boundary. Two sensing methods are commonly used: pulse-echo, where a single transducer both emits and receives the sound wave which is reflected from an interface, and pitch-catch, where one transducer transmits and another receives [8].

Traditional approaches use physical inversion models derived from first principles to convert measured ultrasonic parameters (like velocity or attenuation) into estimates of density, concentration, or other physical attributes [9]. However, industrial processes are often complicated by multiple interfaces and heterogeneities making purely analytical modelling difficult [10, 11]. Therefore, calibration procedures are common which correlate ultrasonic measurements to material properties across a range of process parameters without defining the underlying paths of the sound wave. However, calibration procedures also become complicated in industrial processes when many parameters must be accounted for, such as temperature, gas content, and the content of particulates [12]. Machine learning (ML), particularly supervised ML, offers an alternative to calibration strategies for US sensor measurements [13]. In supervised ML, a model is trained via an algorithm to map input data to labelled data. For example, each US measurement (and any additional process measurements) is paired with a ground truth target taken by a reference measurement. The aim is to produce a model that can predict properties or classes for new, unseen US measurements.

The benefits of applying ML to US measurements compared with calibration procedures include:

1. **Reduced time investment**: Instead of manually calibrating across all parameter ranges, a labelled dataset encompassing natural process variations can be used to train and continuously update an ML model.
2. **Complex waveform features**: ML models can handle input features extracted from the entire ultrasonic waveform rather than just relying on more simple parameters like velocity or attenuation.
3. **Complex mapping**: Compared to most linear or polynomial-based calibration approaches, ML models can fit high-dimensional, non-linear relationships.
4. **Generalisability**: Validation procedures can reduce overfitting, thereby increasing the likelihood that the ML model can handle new data.
5. **Continuous retraining**: As processes evolve or new data arrive, the ML model can be retrained or updated to refine its predictions over time.
6. **Direct correlation to process outcomes**: ML can be applied not only to predict intrinsic material properties (e.g., density) but also to classify process states (e.g., 'fully mixed' or 'fully clean'), or to estimate time remaining until a certain process end point.

4.2 Feature extraction

Despite the advantages listed above, the combination of ML with US sensor measurements requires some domain knowledge to select input features that contain information about the process phenomena being monitored. Empirical testing may be required to plot these features over time and determine whether their variation correspond to process dynamics validated by reference measurements. Feature importance methods, such as Shapley Additive explanations, can also determine the impact of each feature on the prediction being made by an ML model [13]. For US measurements, features can be extracted from:

- **Time-domain waveforms**: This could include direct raw wave amplitudes at each sampled time point, specific amplitudes (maximum, minimum, or peak-to-peak amplitude), amplitude dispersion (standard deviation, spread, entropy), amplitude descent (roll-off, logarithmic decrement, slope), statistical measures (energy, mean, crest factor, skewness, kurtosis), time centre, pulse duration, and TOF.
- **Frequency domain waveforms**: After applying a Fourier transform, features such as the spectral dispersion (standard deviation, smoothness, entropy), rising time, statistical features (energy, crest factor, skewness, kurtosis), average frequency, and bandwidth can be extracted.
- **Time–frequency domain waveforms**: Wavelet transforms (discrete, continuous, or wavelet packet) decompose the signal at different scales and frequencies. By choosing an appropriate mother wavelet (e.g., Daubechies, Symlet, or Morlet), the number of decomposition levels, and the number of vanishing moments, transient features that may be lost in a Fourier analysis can be captured. In practice, the selection of a mother wavelet often visually aligns with the waveform's shape, while the decomposition level and vanishing moments can be optimised within the ML model's validation procedure.
- **Changes between consecutively acquired waveforms**: Time-lagged features, feature gradients, or variations between consecutive waveforms (e.g., the standard deviation of the energy) are crucial for processes that evolve over time (e.g., fermentation, cleaning, flow regime identification), allowing ML models to incorporate historical context to detect trends and temporal variations.
- **Other process features**: Including process-specific parameters such as temperature, mass flow rate, pressure, and material type as supplementary features can enhance the predictive power of ML models as these variables influence the US signal.

4.3 Process monitoring using ultrasonic sensors

Several sensing methods exist for monitoring processes using US measurements. The following section gives an overview of what these sensing methods are and what information contained within US waveforms has been used to detect.

4.4 Single, invasive probes

Single transducers can be used as invasive probes to monitor material properties during processes. For example, Fox *et al* [14] and Salazar *et al* [15] both monitored air incorporation into batters during mixing [14, 15]. Both sensors monitored the changing acoustic impedance of the batter by measuring the reflected sound wave.

Single, invasive probes can also send and receive sound waves after reflection from a surface in a pulse-echo set-up. Hunter *et al* [16] used a multi-frequency acoustic backscatter system to monitor high concentration particle dispersion [16]. Transducers with central frequencies of 1, 2, 4, and 5 MHz were used in pulse-echo mode and the attenuation decay with time of the returning signal was dependent on the particle concentrations. Many works have used reflector plates attached to probes containing transducers. Hoche *et al* [17, 18] used invasive sensor probes with a 50 mm reflector plate transmission distance to monitor alcoholic fermentation in vessels up to 2140 l in size [17]. A sound velocity–density–temperature calibration model was used to predict ethanol concentration during the fermentations. However, this required a secondary measurement of the density of the fermenting wort. Bowler *et al* used ML to predict the alcohol concentration during beer fermentation in a 30 l vessel [18]. The energy of the reflection from the probe–wort interface increased throughout fermentation although no trend in the speed of sound was identified owing to variations in the process temperature. Novoa–Díaz *et al* [19] used an invasive US sensor probe with a reflector plate to monitor malolactic fermentation in red wine [19]. The US velocity increased during the fermentation until reaching a stationary phase after approximately six days. Single invasive sensors with reflector plates have also been used for US attenuation spectroscopy [20, 21].

4.5 Single, non-invasive, through-transmission sensing

Ultrasonic sensors can also be attached externally to vessels enabling non-invasive monitoring. Wallhaußer *et al* [21] combined US measurements and ML to classify whether a model heat exchanger was fouled by dairy protein deposits [22]. A single US sensor monitored waveforms reflected from the plate-fouling interface and from the far wall of the heat exchanger. Úbeda *et al* [22] used a single US transducer to monitor cleaning of milk fouling in a model plate heat exchanger [23]. Two sections of the received US waveforms were analysed: the start of the waveform corresponding to sound wave reflections from the plate-fouling interface, and the third echo reflecting from the far wall of the plate heat exchanger (after passing through the fouling and cleaning fluid or cleaning fluid alone). For the sound wave reflections from the plate-fouling interface, three US waveform features were monitored: the temporal spread, temporal roll-off, and temporal inertia. These features were used in an ML model to classify whether the plate was fouled or cleaned. For the third echo, four features were used: the maximum amplitude, spectral crest factor, spectral centroid, and temporal entropy.

Becker *et al* [23] used a single externally mounted sensor to monitor beer fermentation in a 300 m^3 tank under industrial conditions for 90 h [24]. The US wave was transmitted across the 4.5 m diameter vessel and was reflected at the far wall before returning to the transducer. The US velocity was monitored, and ML was used to compensate for the effects of temperature. Figueiredo *et al* [24] used a single transmission-based US sensor to monitor the transesterification process of biodiesel at small scale (70 mm diameter vessel) [25]. Stabilisation of the US velocity and amplitude indicated the achievement of the maximum yield and that the process should be stopped at this point. Ultrasonic measurements have been widely applied for monitoring membrane fouling during micro-, ultra- and nano-filtration separation processes. The amplitudes of reflected sound waves from the fouling and membrane layers are measured to monitor fouling formation. Li *et al* [25] used an *in situ* US technique to monitor organic and colloidal fouling during nano-filtration [26]. Differences in the fouling processes were observed in the US measurements when using different mixtures of foulants.

4.6 Single, non-invasive, no-transmission sensing

Single, non-invasive US sensors can also be used without transmission through the material of interest. Instead, the sound wave reflection from the interface between the vessel wall and material can be analysed and monitored. This can prevent the need for higher-power transducers to ensure transmission through the system. Escrig *et al* [26] monitored the removal of three types of food fouling (gravy, tomato paste, and malt) at two cleaning fluid temperatures (12 °C and 45 °C) using a single US sensor [27]. The US sensing method monitored the US waveform reflected from the interface between the pipe wall and the fouling material. The pipe section was cuboid with a flat, stainless steel bottom plate where the US sensor was attached externally. Three features from the received US waveform were monitored: the waveform energy, peak-to-peak amplitude, and root mean square error of the amplitudes at every sample point in the waveform compared to that of a clean pipe. Escrig *et al* [11] expanded this study by training classification ML models to predict whether the pipe section was fouled or cleaned [28]. The highest model accuracy was attained when the amplitudes at each sample point in the waveforms were used directly instead of using any further feature extraction methods. Simeone *et al* [10] used the same pipe test section and materials to monitor cleaning by combining US and optical sensors [29]. A three-level wavelet package transform using the 3 Daubechies mother wavelet was performed on the US waveforms. Afterwards, the mean, standard deviation, minimum, maximum, skewness, kurtosis, and energy were extracted from the decomposed signals. These features were then input into an ML model for a regression task to predict the surface area or volume of fouling remaining.

Chen *et al* [27] used a single US sensor to monitor cleaning of wax deposits from a flat duct section over three hours [30]. A decorrelation coefficient of coda waves compared with a clean plate was monitored and cleaning was completed after 2.2 h. Meng *et al* [28] used a single non-invasive sensor to monitor yoghurt fermentation through the wall of a 250 ml stainless steel reactor [31]. The acoustic impedance was monitored using multiple reflections from the wall-yoghurt interface. The acoustic impedance of the yoghurt increased between pH 5.6 and 5.3. Bowler *et al* [29] used single, non-invasive US sensors in reflection-mode to monitor two model mixing systems: honey–water blending and flour–water batter mixing [32]. Classification ML models were developed to predict if the materials were mixed or not, and regression models were trained to predict the time remaining until (or time passed since) the materials were fully mixed.

4.7 Multiple, non-invasive, through-transmission sensing

Many works have attached pairs of non-invasive US sensors to perform through-transmission monitoring of processes in pitch–catch mode. Resa *et al* [30] mounted two non-invasive US sensors to a square glass bottle to monitor the US velocity during alcoholic fermentation [33]. A water bath was used to keep a constant temperature of 30 °C. The US velocity decreased with the decreasing density of the fermenting medium. Resa *et al* [31] used similar experimental methods to monitor lactic acid fermentation [34]. The US velocity decreased throughout fermentation despite no significant change in density. Resa *et al* [12] used the US amplitude to monitor beer wort fermentation and an increase in attenuation was obtained during the start of ethanol production due to the production of CO_2 bubbles [35].

Keskinoğlu and Aydln [32] used two non-invasive transducers in pitch–catch mode to monitor cell growth in a small-scale vessel [36]. From the US velocity, growth curves were obtained that could identify the lag, growth, and stationary phases. Coates *et al* [33] used US and spectroscopic sensors to monitor the blending of polyethylene and polypropylene pellets in a 38 mm single screw extruder [37]. The US velocity was more sensitive than the spectroscopic measurements in detecting a change of blend composition. Barnes *et al* [34] monitored the blending of ethylene vinyl acetate random co-polymers with varying vinyl acetate content [38]. US probes were implemented into dies attached to the end of the extruder barrel and the TOF was measured. Sun *et al* [35] monitored filler dispersion during extrusion using two US sensors [39]. An ML model was trained using the US velocity and attenuation along with the pressure, temperature, filler type and feed rate for prediction of the dispersion index. Halmen *et al* [36] used US tomography to determine filler distribution in polymer melts [40]. The US velocity and attenuation from a 60 mm ID sensor ring consisting of 40 transducers were used.

4.8 Multiple, invasive, through-transmission sensing

Invasive transducer pairs in pitch–catch mode have also been used to reduce the distance needed for sound wave transmission. Ogasawara *et al* [37] used two invasive probes in pitch–catch mode with a 15 mm transmission distance to monitor yoghurt fermentation in a beaker [41]. The US velocity was used to detect phase changes of the yoghurt along with variations in the internal temperature caused by the exothermic fermentation reaction. Mougin *et al* [38] used an invasive probe consisting of two pairs of broadband US transducers to monitor the crystallisation of two organic compounds [42]. US attenuation spectroscopy using the Epstein and Carhart [39] and Allegra and Hawley [40] (ECAH) scattering model was utilised. The frequency range of the probe spanned from 1 MHz to over 150 MHz and was capable of measuring particle sizes in the range 0.01–1000 μm. Bamberger and Greenwood [41] used an invasive probe to monitor slurry suspension in a 1.91 m diameter tank [43]. The US attenuation was measured across three transducer pairs located at different heights along the probe and separated by a distance of 10.2 cm.

4.9 Circulation lines and flow-through cells

Methods to measure smaller material volumes can be used to reduce the sound wave transmission distance, such as positioning US sensors on circulation lines or on dedicated flow-through cells. Hussein *et al* [42] implemented a single US sensor on a circulation line for in-line monitoring of a 60 l yeast fermentation process [44]. The sensor used a reflector plate to transmit through the fermenting liquid. Frequency domain analysis, after Fourier transformation, and phase shift correction were used for the TOF measurements. The US velocity was combined with nine waveform features and the temperature and was inputted into a ML model to predict the liquid density. Mougin *et al* [43, 44] used a 470 ml stainless steel flow-through cell connected to the side-ports of a 2.6 l reactor to determine the crystal size and solid concentration during crystallisation [45]. The method could identify secondary nucleation, growth, and crystal breakage and the results were used to determine kinetic parameters. Tourbin and Frances [45] used a flow-through cell and US attenuation spectroscopy to monitor the suspension and aggregation of nanoparticles in a 1 l stirred tank [46]. Liu *et al* [46] also used a flow-through cell and US attenuation spectroscopy in combination with electrical resistance tomography to monitor crossflow membrane emulsification [47]. The size distribution and concentration of droplets was determined using the ECAH [39] inversion model.

4.10 Future directions: combining machine learning with physics

A particular avenue of research into the analysis of US measurements is the combination of data-driven machine learning with mechanistic equations, often called hybrid modelling. Two developing areas include physics-informed (PINNs) and physics-encoded (PeNNs) neural networks which have been mostly applied to US imaging and non-destructive testing.

PINNs integrate US physics directly within the loss function, enforcing neural network predictions to satisfy known equations or boundary conditions. The network is penalised if it violates the underlying physical equations. For example, PINNs have been used to obtain velocity fields for non-destructive evaluation of plates [47]. Data was collected using laser vibrometers to measure displacement at different spatial locations in the plate. However, it was posited that sound velocity would be lower in areas where a flaw was located. The wave equation was utilised to convert the displacement data to velocity. Two neural networks were trained at the same time, one to predict displacement and the other to predict sound velocity. The networks were trained on a combined loss from the displacement data and the wave equation [48]. Similarly, PINNs have been used to find the spatial variation of elastic coefficients in materials by using elastic wave equations [48] and intraventricular vector flow mapping [49]. Direct integration of physical equations within ML can reduce data requirements while increasing accuracy and generalisability.

On the other hand, PeNNs structurally encode US physics directly into neural network architectures. For example, traditional iterative solvers for US beamforming and image reconstruction alternate between minimising a data-fidelity term and applying a regularisation operator. Deep unfolding unrolls this algorithm into a fixed number of stages, making each iteration a layer in a convolutional neural network [50]. Therefore, iterative solver methods can be embedded into convolutional neural networks (e.g., by adding denoisers, learned thresholding, or adaptive regularisation blocks) so that the model can be trained via backpropagation. The result is an interpretable network structure that mirrors the iterations of an established algorithm. Partial differential equations (PDEs) for wave equations have also been solved using PeNNs [51]. Partial differential equations are typically solved using finite differences, finite elements, or other mesh-based methods which can become computationally expensive. Convolutional layers can be used to replicate PDE time or spatial stepping. Therefore, the PDE can be hard-wired into the architecture of the neural network.

References

[1] Ghobakhloo M 2020 Industry 4.0, digitization, and opportunities for sustainability *J. Clean. Prod.* **252** 119869

[2] Kang H S, Lee J Y, Choi S, Kim H, Park J H, Son J Y *et al* 2016 Smart manufacturing: past research, present findings, and future directions *Int. J. Pr. Eng. Manuf.—Green Technol.* **3** 111–28

[3] De Beer T, Burggraeve A, Fonteyne M, Saerens L, Remon J P and Vervaet C 2011 Near infrared and Raman spectroscopy for the in-process monitoring of pharmaceutical production processes *Int. J. Pharm.* **417** 32–47

[4] Sjödin D, Parida V, Leksell M and Petrovic A 2018 Smart factory implementation and process innovation: a preliminary maturity model for leveraging digitalization in manufacturing *Res. Technol. Manag.* **61** 22–31

[5] Fisher O, Watson N, Porcu L, Bacon D, Rigley M and Gomes R L 2018 Cloud manufacturing as a sustainable process manufacturing route *J. Manuf. Syst.* **47** 53–68

[6] Henning B and Rautenberg J 2006 Process monitoring using ultrasonic sensor systems *Ultrasonics* **44** e1395–e9

[7] McClements D J 1995 Advances in the application of ultrasound in food analysis and processing *Trends Food Sci. Technol.* **6** 293–9

[8] Awad T S, Moharram H A, Shaltout O E, Asker D and Youssef M M 2012 Applications of ultrasound in analysis, processing and quality control of food: a review *Food Res. Int.* **48** 410–27

[9] Challis R E, Povey M J W, Mather M L and Holmes A K 2005 Ultrasound techniques for characterizing colloidal dispersions *Rep. Prog. Phys.* **68** 1541–637

[10] Simeone A, Woolley E, Escrig J and Watson N J 2020 Intelligent industrial cleaning: a multi-sensor approach utilising machine learning-based regression *Sensors (Basel)* **20** 3642

[11] Escrig J E, Simeone A, Woolley E, Rangappa S, Rady A and Watson N J 2020 Ultrasonic measurements and machine learning for monitoring the removal of surface fouling during clean-in-place processes *Food Bioprod. Process.* **123** 1–13

[12] Resa P, Elvira L, Montero de Espinosa F, González R and Barcenilla J 2009 On-line ultrasonic velocity monitoring of alcoholic fermentation kinetics *Bioprocess. Biosyst. Eng.* **32** 321–31
[13] Bowler A L, Pound M P and Watson N J 2022 A review of ultrasonic sensing and machine learning methods to monitor industrial processes *Ultrasonics* **124** 106776
[14] Fox P, Smith P P and Sahi S 2004 Ultrasound measurements to monitor the specific gravity of food batters *J. Food Eng.* **65** 317–24
[15] Salazar J, Turó A, Chávez J A and García M J 2004 Ultrasonic inspection of batters for on-line process monitoring *Ultrasonics* **42** 155–9
[16] Hunter T N, Darlison L, Peakall J and Biggs S 2012 Using a multi-frequency acoustic backscatter system as an *in situ* high concentration dispersion monitor *Chem. Eng. Sci.* **80** 409–18
[17] Hoche S, Krause D, Hussein M A and Becker T 2016 Ultrasound-based, in-line monitoring of anaerobe yeast fermentation: model, sensor design and process application *Int. J. Food Sci. Technol.* **51** 710–9
[18] Bowler A, Escrig J, Pound M and Watson N 2021 Predicting alcohol concentration during beer fermentation using ultrasonic measurements and machine learning *Fermentation* **7** 34
[19] Novoa-Díaz D, Rodríguez-Nogales J M, Fernández-Fernández E, Vila-Crespo J, García-Álvarez J, Amer M A *et al* 2014 Ultrasonic monitoring of malolactic fermentation in red wines *Ultrasonics* **54** 1575–80
[20] Shukla A, Prakash A and Rohani S 2010 Online measurement of particle size distribution during crystallization using ultrasonic spectroscopy *Chem. Eng. Sci.* **65** 3072–9
[21] Wallhäußer E, Hussein W B, Hussein M A, Hinrichs J and Becker T M 2011 On the usage of acoustic properties combined with an artificial neural network—a new approach of determining presence of dairy fouling *J. Food Eng.* **103** 449–56
[22] Úbeda M A, Hussein W B, Hussein M A, Hinrichs J and Becker T M 2016 Acoustic sensing and signal processing techniques for monitoring milk fouling cleaning operations *Eng. Life Sci.* **16** 67–77
[23] Becker T, Mitzscherling M and Delgado A 2002 Hybrid data model for the improvement of an ultrasonic-based gravity measurement system *Food Control* **13** 223–33
[24] Figueiredo M K K, Silva C E R, Alvarenga A V and Costa-Félix R P B 2018 Relating speed of sound and echo amplitude with biodiesel manufacture *Chem. Eng. Res. Des.* **136** 825–33
[25] Li X, Zhang H, Hou Y, Gao Y, Li J, Guo W and Ngo H H 2015 *In situ* investigation of combined organic and colloidal fouling for nanofiltration membrane using ultrasonic time domain reflectometry *Desalination* **362** 43–51
[26] Escrig J, Woolley E, Rangappa S, Simeone A and Watson N J 2019 Clean-in-place monitoring of different food fouling materials using ultrasonic measurements *Food Control.* **104** 358–66
[27] Chen B, Callens D, Campistron P, Moulin E, Debreyne P and Delaplace G 2019 Monitoring cleaning cycles of fouled ducts using ultrasonic coda wave interferometry (CWI) *Ultrasonics* **96** 253–60
[28] Meng R F, Zhou J W, Ye X Q and Liu D H 2012 On-line monitoring of yogurt fermentation using acoustic impedance method *Appl. Mech. Mater.* **101–102** 737–42
[29] Bowler A L, Bakalis S and Watson N J 2020 Monitoring mixing processes using ultrasonic sensors and machine learning *Sensors* **20** 1813

[30] Resa P, Elvira L and Montero de Espinosa F 2004 Concentration control in alcoholic fermentation processes from ultrasonic velocity measurements *Food Res. Int.* **37** 587–94
[31] Resa P, Elvira L, Espinosa F M, Bolumar T and Perez G 2004 Ultrasonic velocity measurements in the ternary mixtures water-lactose-lactate, for the purpose of monitoring the lactic acid fermentation of lactose *IEEE Ultrasonics Symp., 2004 (23–27 Aug. 2004)*
[32] Keskinoğlu C and Aydın A 2021 Ultrasound based noninvasive real-time cell proliferation process monitoring *J. Acoust. Soc. Am.* **149** 3345
[33] Coates P D, Barnes S E, Sibley M G, Brown E C, Edwards H G M and Scowen I J 2003 In-process vibrational spectroscopy and ultrasound measurements in polymer melt extrusion *Polymer* **44** 5937–49
[34] Barnes S E, Brown E C, Sibley M G, Edwards H G, Scowen I J and Coates P D 2005 Vibrational spectroscopic and ultrasound analysis for in-process characterization of high-density polyethylene/polypropylene blends during melt extrusion *Appl. Spectrosc.* **59** 611–9
[35] Sun Z, Jen C-K, Yan J and Chen M-Y 2005 Application of ultrasound and neural networks in the determination of filler dispersion during polymer extrusion processes *Polym. Eng. Sci.* **45** 764–72
[36] Halmen N, Kugler C, Hochrein T, Heidemeyer P and Bastian M 2017 Ultrasound tomography for inline monitoring of plastic melts *J. Sens. Sens. Syst.* **6** 9–18
[37] Ogasawara H, Mizutani K, Ohbuchi T and Nakamura T 2006 Acoustical experiment of yogurt fermentation process *Ultrasonics* **44** e727–e30
[38] Mougin P, Wilkinson D, Roberts K J and Tweedie R 2002 Characterization of particle size and its distribution during the crystallization of organic fine chemical products as measured *in situ* using ultrasonic attenuation spectroscopy *J. Acoust. Soc. Am.* **109** 274–82
[39] Epstein P S and Carhart R R 1953 The absorption of sound in suspensions and emulsions. I. Water fog in air *J. Acoust. Soc. Am.* **25** 553–65
[40] Allegra J R and Hawley S A 1972 Attenuation of sound in suspensions and emulsions: theory and experiments *J. Acoust. Soc. Am.* **51** 1545–64
[41] Bamberger J A and Greenwood M S 2004 Using ultrasonic attenuation to monitor slurry mixing in real time *Ultrasonics* **42** 145–8
[42] Hussein W B, Hussein M A and Becker T 2012 Robust spectral estimation for speed of sound with phase shift correction applied online in yeast fermentation processes *Eng. Life Sci.* **12** 603–14
[43] Mougin P, Wilkinson D and Roberts K J 2003 *In situ* ultrasonic attenuation spectroscopic study of the dynamic evolution of particle size during solution-phase crystallization of urea *Cryst. Growth Des.* **3** 67–72
[44] Mougin P, Thomas A, Wilkinson D, White G, Roberts K J, Herrmann N *et al* 2003 On-line monitoring of a crystallization process *AIChE J.* **49** 373–8
[45] Tourbin M and Frances C 2009 Monitoring of the aggregation process of dense colloidal silica suspensions in a stirred tank by acoustic spectroscopy *Powder Technol.* **190** 25–30
[46] Liu L, Li R F, Collins S, Wang X Z, Tweedie R and Primrose K 2011 Ultrasound spectroscopy and electrical resistance tomography for online characterisation of concentrated emulsions in crossflow membrane emulsifications *Powder Technol.* **213** 123–31
[47] Shukla K, Di Leoni P C, Blackshire J, Sparkman D and Karniadakis G E 2020 Physics-informed neural network for ultrasound nondestructive quantification of surface breaking cracks *J. Nondestrct. Eval.* **39** 61

[48] Shukla K, Jagtap A D, Blackshire J L, Sparkman D and Karniadakis G E 2022 A physics-informed neural network for quantifying the microstructural properties of polycrystalline nickel using ultrasound data: a promising approach for solving inverse problems *IEEE Signal Process. Mag.* **39** 68–77

[49] Ling H J, Bru S, Puig J, Vixège F, Mendez S, Nicoud F *et al* 2024 Physics-guided neural networks for intraventricular vector flow mapping *IEEE Trans. Ultrason. Ferroelectr. Freq. Control* **71** 1377–88

[50] Luijten B, Chennakeshava N, Eldar Y C, Mischi M and van Sloun R J G 2023 Ultrasound signal processing: from models to deep learning *Ultrasound Med. Biol.* **49** 677–98

[51] Mehtaj N and Banerjee S 2025 Scientific machine learning for guided wave and surface acoustic ave (SAW) propagation: PgNN, PeNN, PINN, and Neural Operator *Sensors* **25** 1401

Chapter 5

Exemplars of optical and acoustic measurements including tomography—MJP

Megan J Povey

Acoustic waves of all frequencies respond to the thermo-physical properties of materials, in particular thermal diffusivity, thermal conductivity, heat capacity, density, adiabatic compressibility (except at GHz frequencies where properties tend to isothermal) and their variation across interfaces. On the other hand, optical and electromagnetic waves respond to dielectric factors, magnetic permeability, and electrical conductivity with density the only property in common with the thermal. As a result there is enormous potential in combining the two modalities into a versatile tomographic tool.

5.1 Example 1: near infrared reflectance spectroscopy and ultrasound velocity measurement characterising the brew time of green tea: an illustration of the buffer rod technique for high resolution ultrasound spectroscopy

Dragon Well Tea is a notable Chinese green tea variety. The release and concentration of bioactive compounds, such as catechins and caffeine, which have antioxidant, anticancer, and antidiabetic properties, are significantly influenced by brewing [1]. Optimal brewing enhances these benefits, whilst methods such as taste panels are subjective, time consuming, and destructive. This study employs non-destructive techniques—sensory analysis, ultrasound technology, near infrared reflectance (NIR) spectroscopy—to provide comprehensive data on the tea's physical and chemical changes during brewing.

5.1.1 Method

Details of sample preparation and equipment are given in figures 5.1–5.5.

doi:10.1088/978-0-7503-3203-3ch5

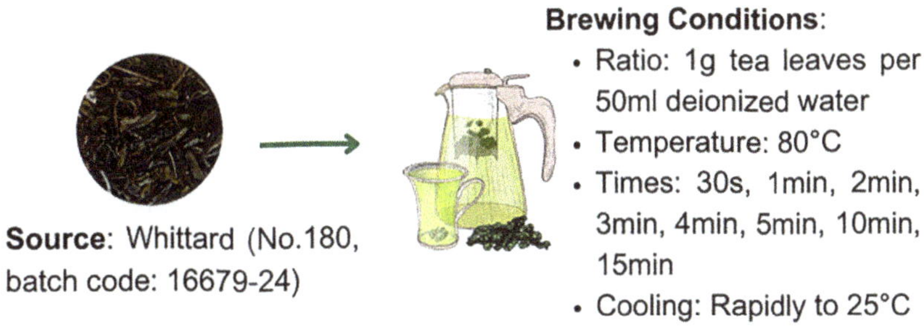

Figure 5.1. Sample preparation.

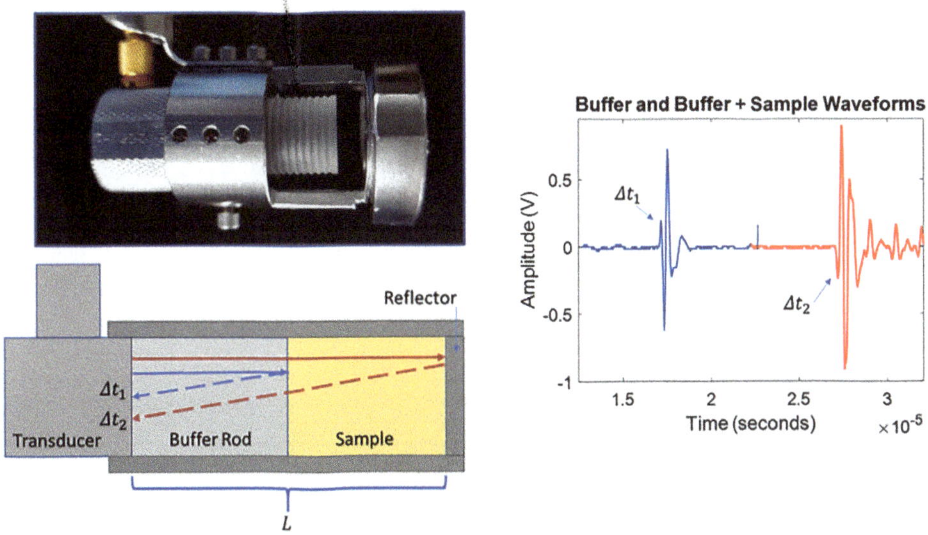

Figure 5.2. Schematic of the acoustic probe (left) and the measured waveforms obtained from the buffer rod reflection (blue) and the buffer rod together with the sample (red—right). Δt1 and Δt2 represent the time-of-flight of the pulse travelling through the buffer rod and through the buffer rod and sample, respectively [2].

5.1.2 Measurements

Five trained evaluators GBT23776-2018 Total Score = Leaf Bottom × 10% + Liquor Color × 20% + Aroma × 35% + Taste × 35%.

Bottom, Liquor Color, Aroma, and Taste scores increase steadily from 30 s to 3 min, peaking at the 3 min mark. After 3 min, the scores for all parameters begin to decline gradually (figure 5.6).

High resolution speed of sound measurements on the brewing of tea (figure 5.7) shows the changes in the ultrasound parameter (Δx in ns) over the brewing time of 0–15 min. The parameter Δx represents the difference in the sound speed between

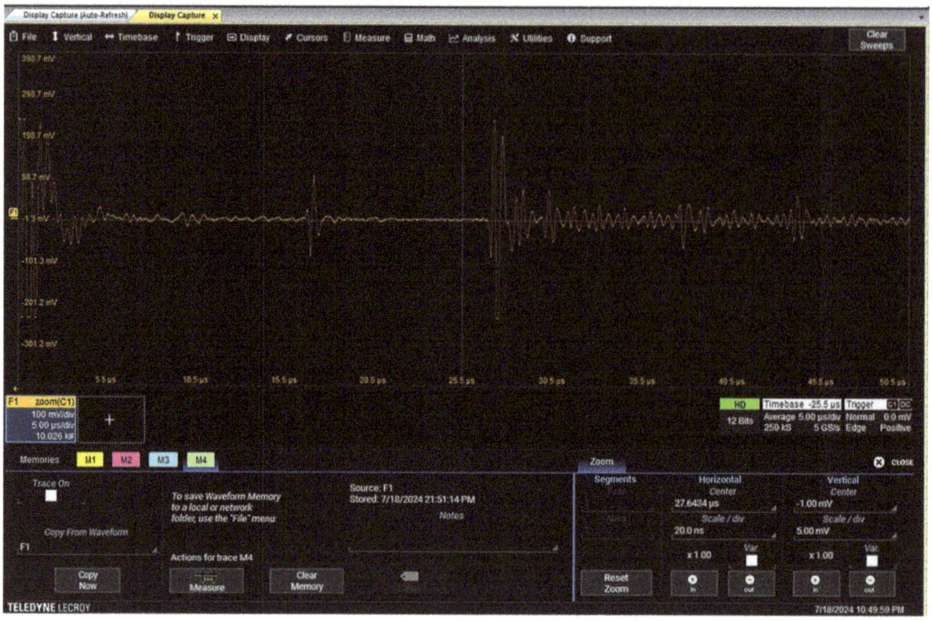

Figure 5.3. Digital oscilloscope view of the reflection at the buffer rod sample interface and the first reflection from the far wall reflector showing the signal that has travelled both through the buffer rod and the sample and back again. Note the lengthy reverberation following the initial pulse, this is called the coda wave and arises due to multiple reflections and paths followed by the sound wave.

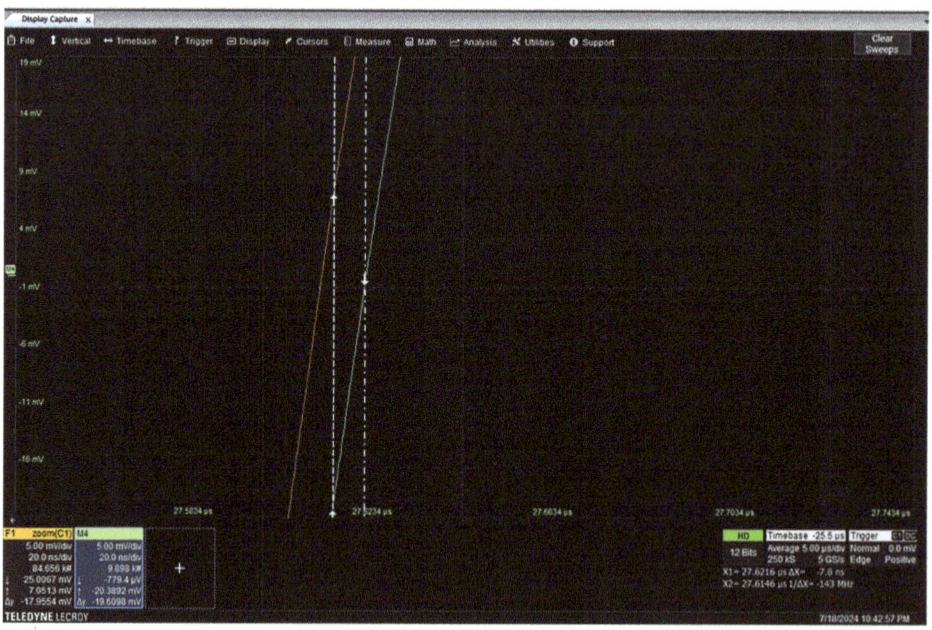

Figure 5.4. Zoom of the first rising portion of the buffer-sample pulse, first in deionised water and then in the tea sample. In this example the difference in arrival time is 7.0 ns.

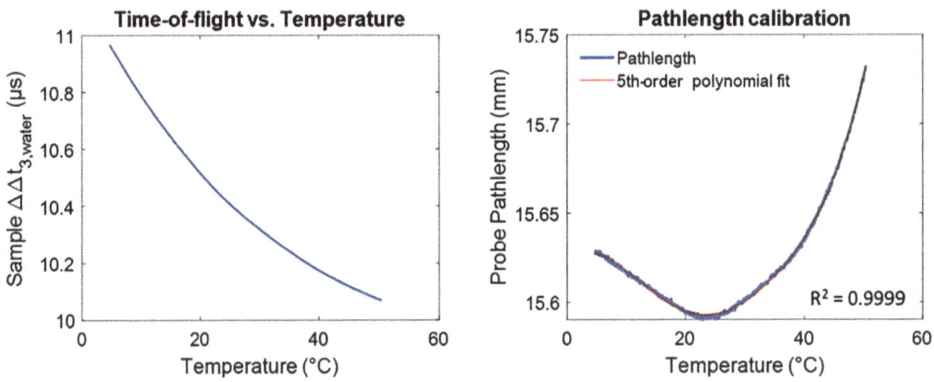

Figure 5.5. Calibration of the sample path length determined as a function of temperature.

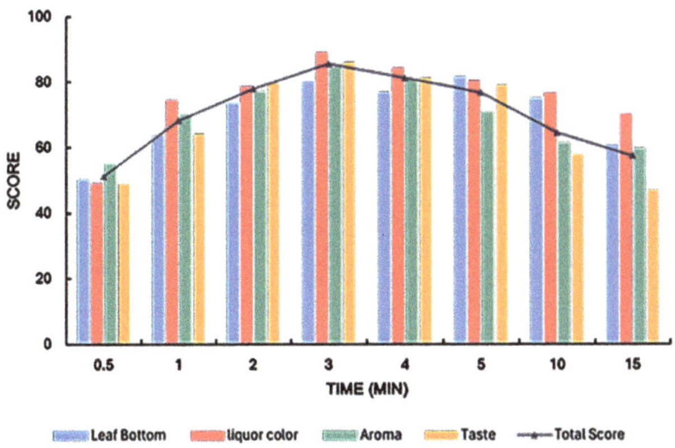

- **Leaf Bottom, Liquor Color, Aroma, and Taste scores increase steadily from 30 seconds to 3 minutes, peaking at the 3-minute mark.**
- **After 3 minutes, the scores for all parameters begin to decline gradually.**

Figure 5.6. Sensory evaluation scores of Dragon Well tea at various brewing times.

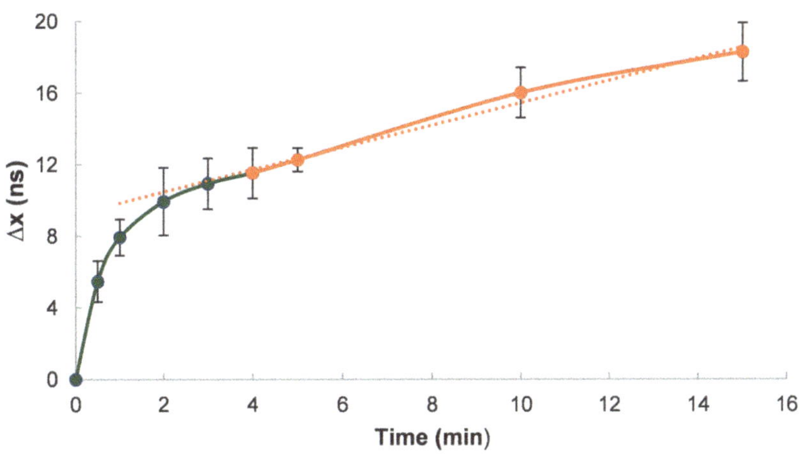

Figure 5.7. Changes in the ultrasound parameter (Δx in ns) over the brewing time from 0 to 15 min. The parameter Δx represents the difference in the sound speed between the tea sample and deionized water. The change from a non-linear to a linear dependence on time matches the ideal brewing time determined by the taste panel.

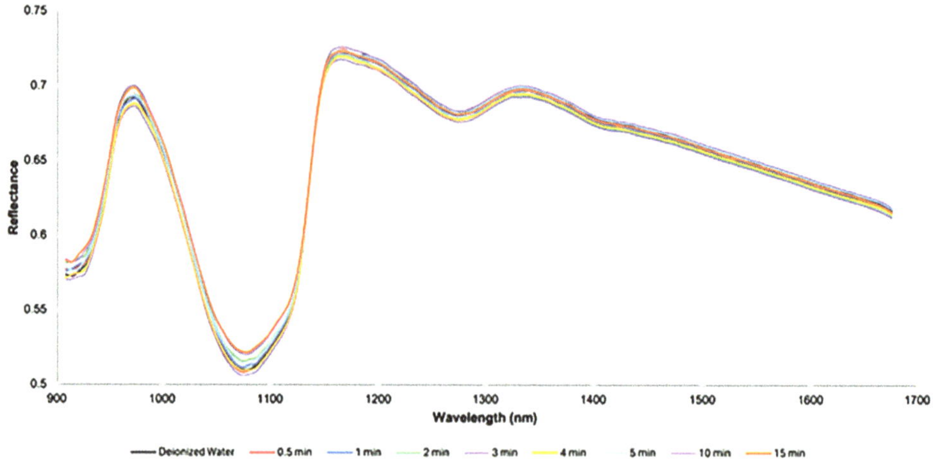

Figure 5.8. NIR reflectance spectra of Dragon Well Tea plotted against brewing time.

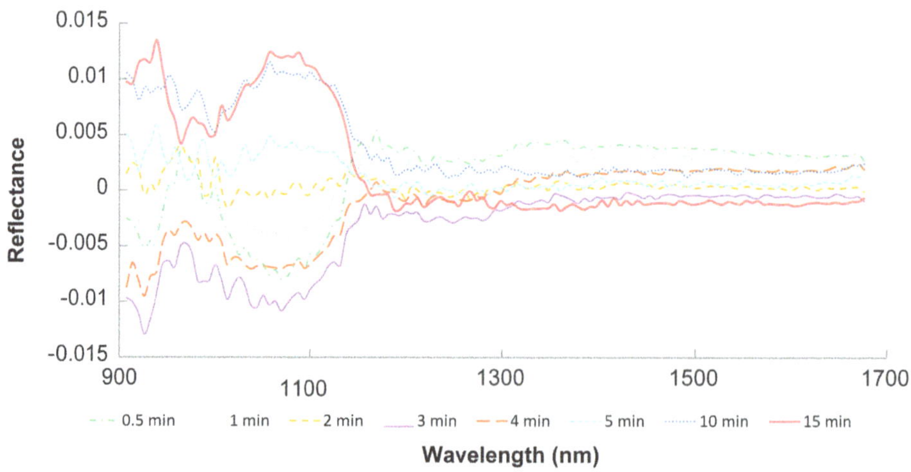

Figure 5.9. Differential NIR reflectance spectra of Dragon Well Tea relative to deionised water.

the tea sample and deionized water, as measured by high resolution ultrasound spectroscopy (figure 5.8).

In figure 5.8 note that spectral differences are difficult to discern because the primary source of reflection is water. Once the water signal is removed, the differences are apparent (figures 5.9 and 5.10).

This work was carried out as part of a Leeds MSc project by Yi An under my supervision.

5.2 Example 2: passive acoustic monitoring

Passive acoustic monitoring has become a standard technique for the assessment of texture in foods. Povey *et al* established the link between crack propagation and

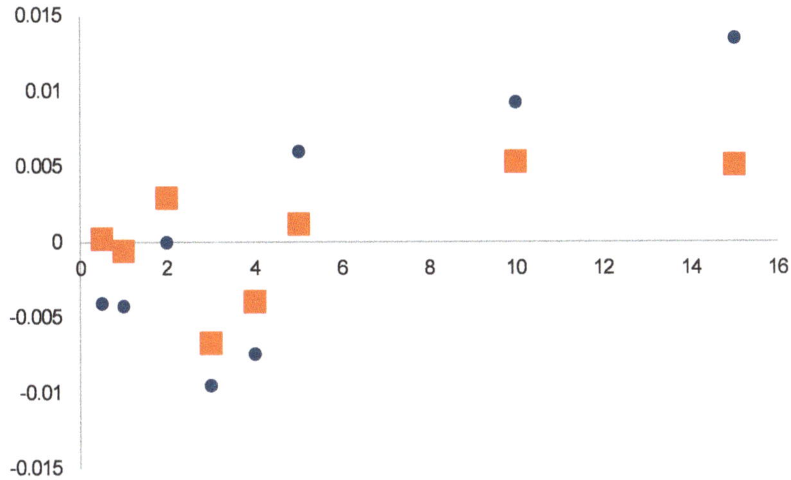

Figure 5.10. Relative reflectance plotted against time in minutes at (orange filled squares) at 1001 nm and (blue filled circles) 939 nm. Note the negative reflectance (absorption peak) at 3 min.

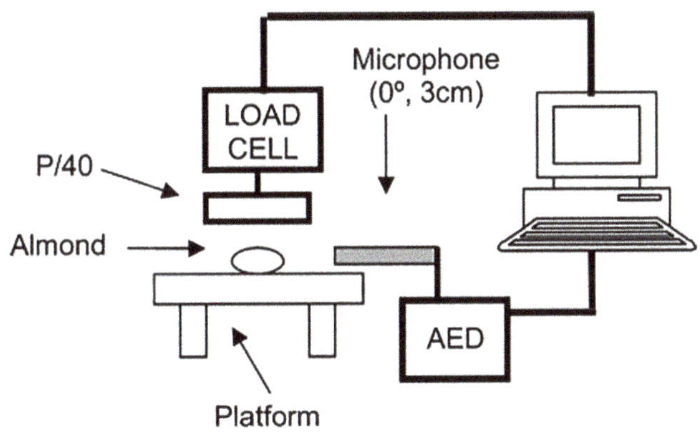

Figure 5.11. Diagram of the instrument setup for the acoustic assessment of texture.

crispness/crunchiness in foods [3] and later in 2005 Varela *et al* demonstrated how passive acoustics could be related to sensory and structural attributes of a food (in this case almonds) [4].

A TA-XT plus Texture Analyser (Stable Micro Systems, Godalming, UK) was used for force/displacement measurement with a 30 kg load cell, together with a P/40 cylinder (40 mm diameter) used for the compression tests. The test settings were: test speed 1 mm s^{-1}, target mode: strain, strain 50%, trigger 10 g. An AED [5] was used for sound recording together with the corresponding software (Texture Exponent32). The gain of the AED was set at one. A Bruel and Kjaer free-field microphone (diameter 8 mm), calibrated using an Acoustic Calibrator Type 4231 (94 and 114 dB SPL 1000 Hz) was positioned at 3 cm distance and with an angle of 8° to the sample (figure 5.11). This angle minimised the distortion that the probe created in the sound field. A high pass filter of 1 kHz removed extraneous ambient acoustic and mechanical noise. A low pass filter set the upper calibrated and measured frequency at 16 kHz. The AED operates by integrating all the frequencies within the band pass range generating a voltage proportional to the SPL. The integration time was set at 0.25 ms. The data acquisition rate was 500 points per second for both force and acoustic signals. All tests were performed in a laboratory with no special soundproof facilities, with a relative humidity of around 25%, and a temperature of 22°C–28°C. Each test was performed on six almonds from each batch (the raw one and the four different roasting times—six replications on each sample). Force versus displacement and SPL versus displacement were plotted, the parameters extracted from the curves.

Varela *et al* [4] demonstrated a very high correlation between the human sensory attributes of crispness/crunchiness and their objective acoustic measures in terms of the rate of generation of the peaks in figure 5.12. later, Gregersen *et al* [6, 7] applied passive acoustics to the characterisation of crystallised fat.

Hefft, Povey, and Zhang [8] later established a unified theoretical–experimental framework for quantifying the sudden energy releases that occur during frictional interfaces and linking them to acoustic emissions (AE). By solving heat-conduction equations under sliding-contact boundary conditions and capturing *in situ* ultrasound signals, the authors identify and interpret three distinct AE regimes—micro-cracks, asperity impacts, and stick-slip chirps—each tied to unique interfacial mechanics.

Frictional heating at a sliding interface converts mechanical work into thermal energy. For two flat surfaces pressed together by a normal force F_n and sliding at velocity (v), the local heat flux is

$$Q = \int_0^t \sigma v \Delta T dt$$

where the stress σ in the contact region whose true area of contact between the surfaces is A, is given by

$$\sigma = \frac{F_n}{A} \tag{5.1}$$

The transient temperature rise ΔT represents the time-dependent increase in temperature due to frictional heating. By integrating the local heat generation over space and time, the cumulative energy release is given by

$$Q = \sigma v \int_0^t \Delta T\left(1 - \mathrm{erf}\left(\frac{x}{2\sqrt{\alpha t}}\right)\right)dx \tag{5.2}$$

Were erf is the error function and the thermal diffusivity $\alpha = \frac{k}{\rho C p}$

5.2.1 Experimental AE validation

Using a pin-on-disc tribometer equipped with a broadband AE sensor and wavelet-based denoising (MATLAB R2020B), the authors recorded time-frequency AE maps during sliding (figure 5.13).

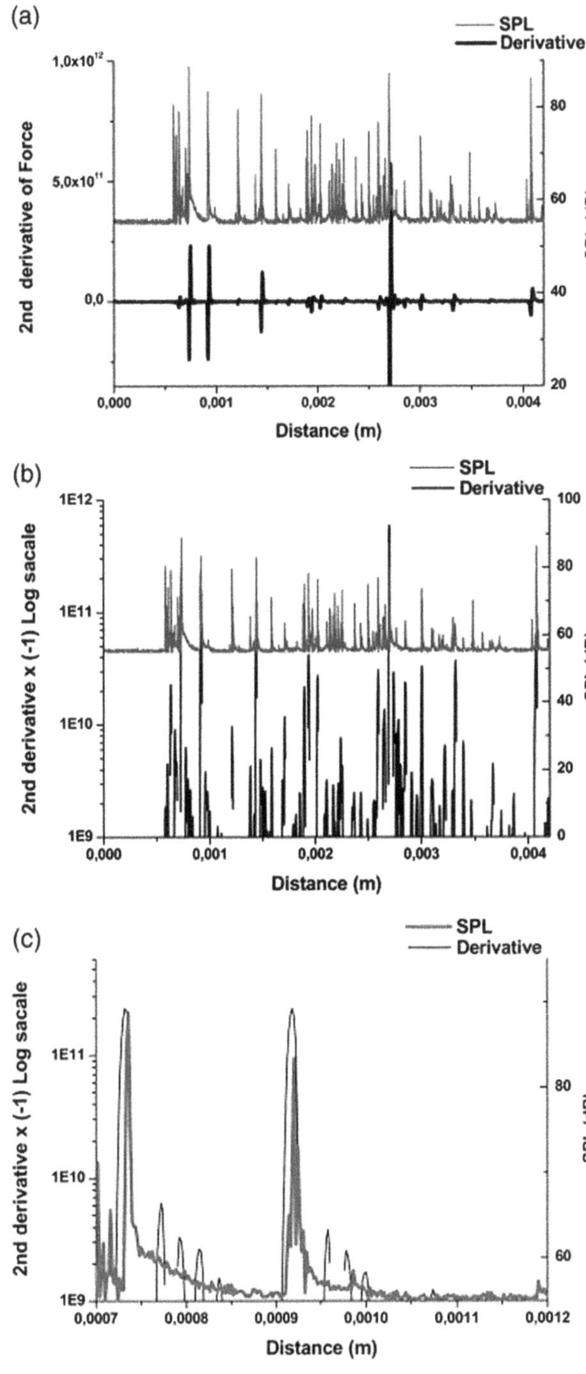

Figure 5.12. Correspondence of force (black lines) and SPL plots (grey lines), example for the 6 min roasted sample, (a) using the second derivative of the force curve, (b) using the negative values of the second derivative of the force curve, (c) enlarged section of (b), correspondence of the shapes of the force and SPL plots, using the negative values of the second derivative.

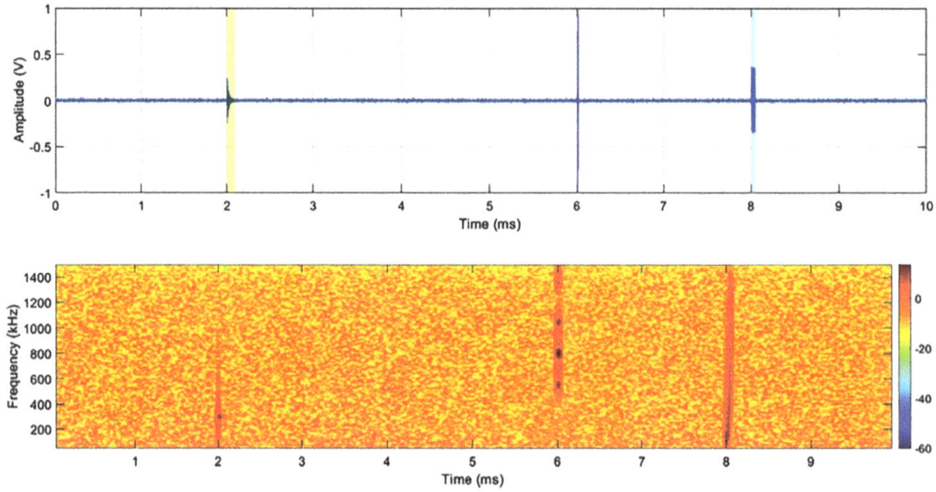

Figure 5.13. Acoustic emission waveform in its time domain (top) and its time–frequency domain (bottom). Yellow highlight = Micro-crack event showing 300 kHz dominant frequency,

Three regimes emerge (figure 5.14):

a) **Yellow: Micro-cracks (300 kHz)** Low-frequency, longer-duration bursts linked to subsurface fracture and thermal diffusion timescales.
b) **Purple: Asperity Impacts (800 kHz and 1.6 MHz)** Harmonic-rich spikes from elastic collisions of surface asperities, where non-linear contact stiffness generates harmonics.
c) **Light Blue: Stick-Slip Chirps (50–600 kHz)** Broadband frequency sweeps whose instantaneous rate tracks interfacial shear-stress buildup and slip rupture.

5.2.2 Lubricated contact effects

Introducing a lubricant modifies both friction and AE:

- **Viscous Damping** Smooths pressure peaks and suppresses high-frequency AE.
- **Film Thickness** Alters shear-rate distributions and heat-generation profiles.
- **Rheology** Non-Newtonian or temperature-dependent viscosities shift predicted AE signatures.

The coupled solid–lubricant model underscores limitations in assuming constant properties and linear viscous behaviour.

5.2.3 Practical implications and limitations

Key assumptions—homogeneous contact stresses, temperature-independent material properties, and Gaussian roughness—warrant extensions using stochastic contact models and high-strain-rate plasticity for nano-confined and non-Newtonian films.

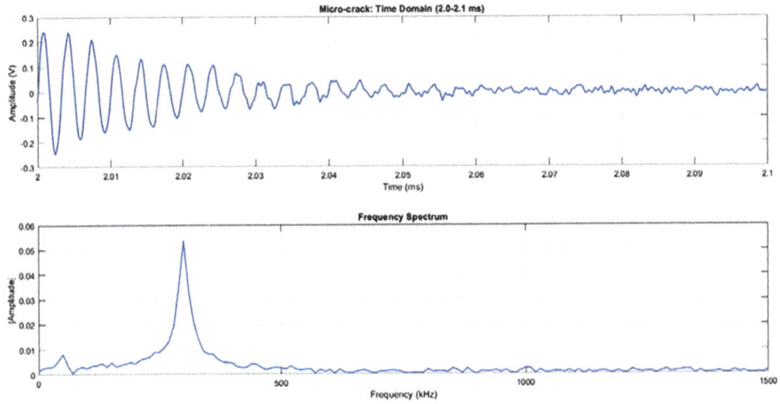

(a) Detailed view of the micro-crack event.

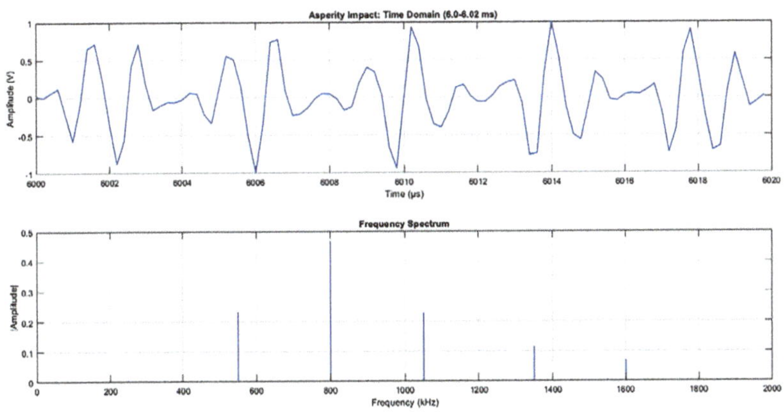

(b) Detailed view of the asperity impact event.

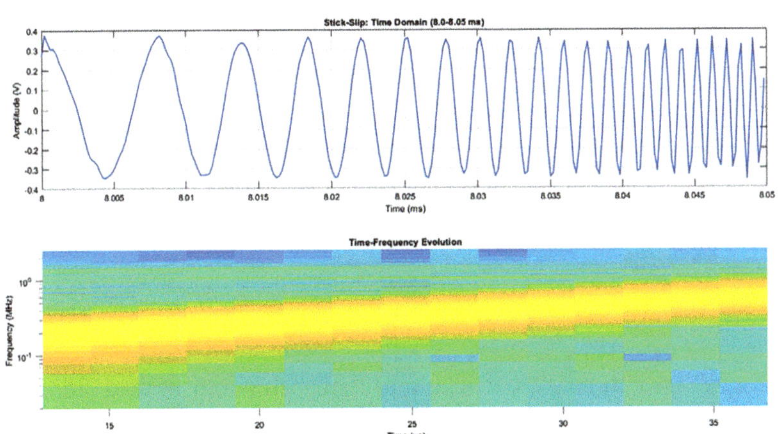

(c) Detailed view of the stick-slip event.

Figure 5.14. Details of events (a–c) in figure 5.13 Acoustic emission waveform in its time domain (top) and its time–frequency domain (bottom). Yellow highlight = Micro-crack event showing 300 kHz dominant frequency.

- **AE-Based Wear Monitoring** Early detection of 300 kHz micro-cracks can signal subsurface fatigue before visible damage.
- **Lubricant Design** Tuning lubricant rheology to attenuate 1.6 MHz harmonics may reduce damaging impacts.

5.2.4 Conclusions

By marrying analytical heat-transfer solutions with *in situ* AE diagnostics, this study quantifies how distinct tribological events manifest as characteristic transient energy releases. The error-function temperature profile predicts interfacial heating, while AE regimes map to micro-mechanisms: fracture, elastic rebound, and frictional instability. This work enables two advances in tribological engineering:

- **AE-based wear monitoring** The measured 300 kHz micro-crack signature figure figure could provide an early warning indicator for subsurface fatigue, detectable prior to macroscopic wear scar formation.
- **Lubricant optimization** By correlating the dimensionless pressure ratio PVPs (equations (14), (16)) in [8] with AE harmonic attenuation, lubricants can be formulated to suppress damaging asperity rebounds (event observed at 1.6 MHz). Overall, this study establishes transient energy release as a useful technique for analyzing tribological systems. By unifying frictional heating models, lubricant rheology, and AE diagnostics, it provides a methodology to dissect interfacial energy dissipation mechanisms across scales.

5.3 Example 3: the Loughborough–Leeds ultrasound diffractometer

We are interested in the scattering behaviour of particles in liquids—for characterisation purposes. Particularly micron-sized particles, typically MHz frequency, linear ultrasound and working in the long wavelength region.

The scattering from a single spherical particle has an angular dependence caused by the contrast in particle properties with surrounding liquid.

Dominant modes are monopole (compressibility and thermal contrast) and dipole (density contrast and viscous limited).

The far field amplitude as a function of angle θ is given by

$$f(\theta) = \frac{1}{ik}\sum\nolimits_n (2n + 1)A_n P_n(\cos\theta) \tag{5.3}$$

where $ka \ll 1$.

The wave vector $k = \frac{2\pi}{\lambda}$ where λ is wavelength, A_n and P_n are the scattering coefficients. (figure 5.15)

But we can't look at single particles and their behaviour in field, and actually we are more interested in the particle systems, not isolated particles.

The diffractometer uses the scattering of sound from silicone tubing, which is filled with water and then a sample which could be as little as 0.3 ml can be pumped into the tube and the change in the angular dependence of the scattering observed. The acoustic impedance of the silicone tubing is not very different to that of water,

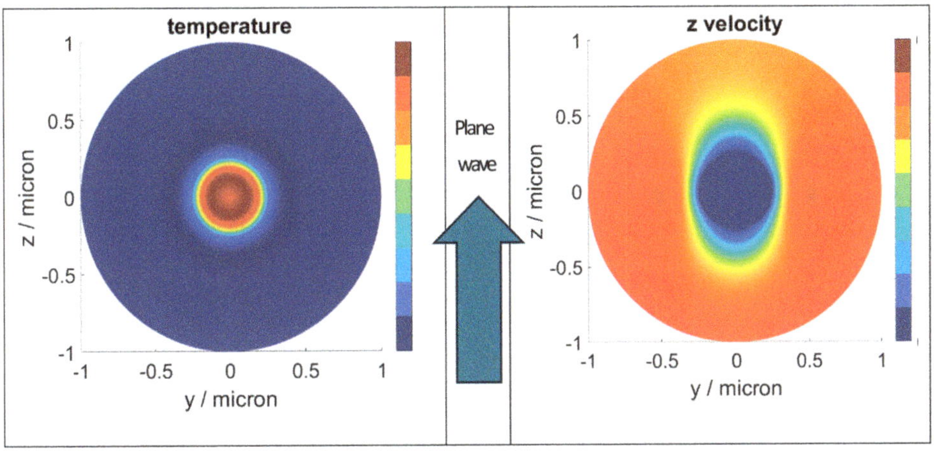

Figure 5.15. Scattering patterns for left: monopole-dominated liquid particles in liquid; and right: dipole-dominated solid particles suspended in liquid.

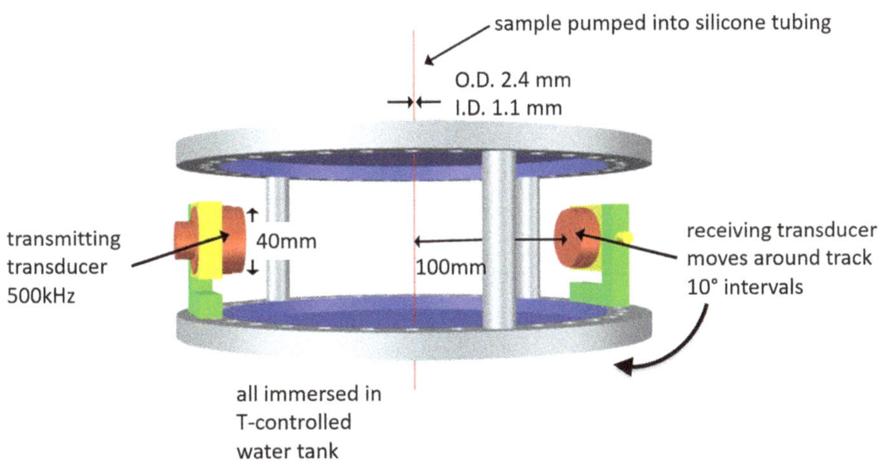

Figure 5.16. Diagram of Leeds–Loughborough diffractometer.

minimizing the scattering from the tubing and permitting observation of scattering from the tube contents (figures 5.16, 5.17).

The effective properties $K_{\mathrm{eff}} < =f(\theta)$ of ensembles of particles are dependent on $f(\theta)$. The angular dependence of a cylindrical sample of effective medium relates to effective properties and single particle far field angular dependence. The clustering of particles also modifies the effective properties through particle interactions.

To interpret the observed scattering data we first analyse the scattering of a plane wave by a single core–shell cylinder in 2D.

Figure 5.19 shows the received signal in response to a single cycle input (figure 5.18) for water filling the silicone tube. The analytical model prediction is

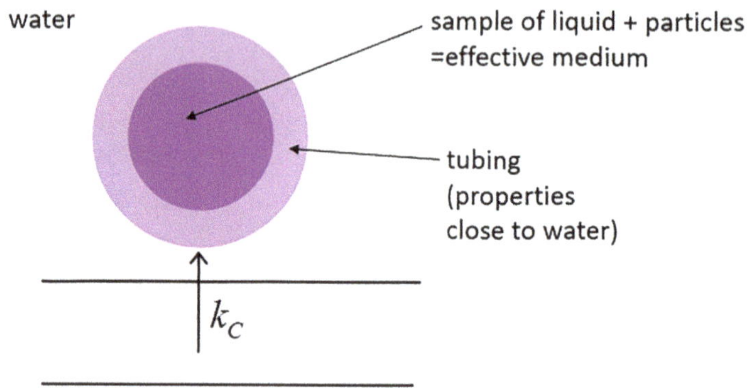

Figure 5.17. Scattering of a plane wave by a single core–shell cylinder in 2D in the long (acoustic) wave limit for particles $k_c a \ll 1$ where a is particle radius. The silicone tubing id is 1 mm.

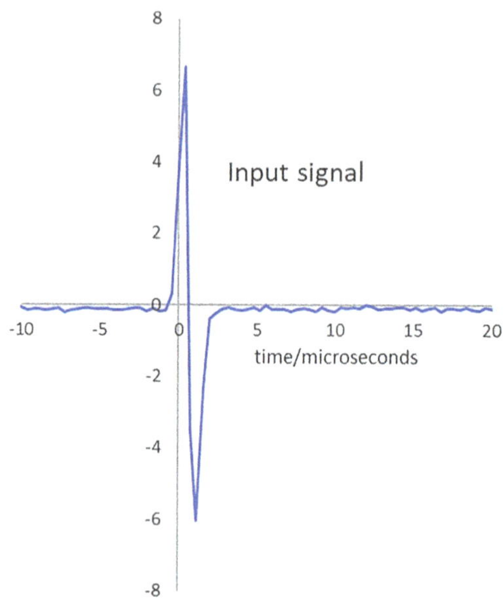

Figure 5.18. Time domain plot of a single cycle input signal

good for arrival time and overall shape but there is some variation between repeats and there are some significant differences in shape (figure 5.20).

Examination of the figure indicates that the angular dependence of the scattering from the skimmed and whole milk samples are similar to that of water. These are preliminary data from an EPSRC sponsored project 'EP/M026302/1 Ultrasonic propagation in complex media: correlated spatial distributions and multiple dispersed phases' led by Professor Valerie Pinfield (Loughborough), Dr Artur Gower and Professor William Parnell (Manchester University), Dr Michael Forrester (Loughborough), Professor David Abrahams (Cambridge), Dr Melvin Holmes, Dr Matt Francis and Professor Megan Povey (Leeds).

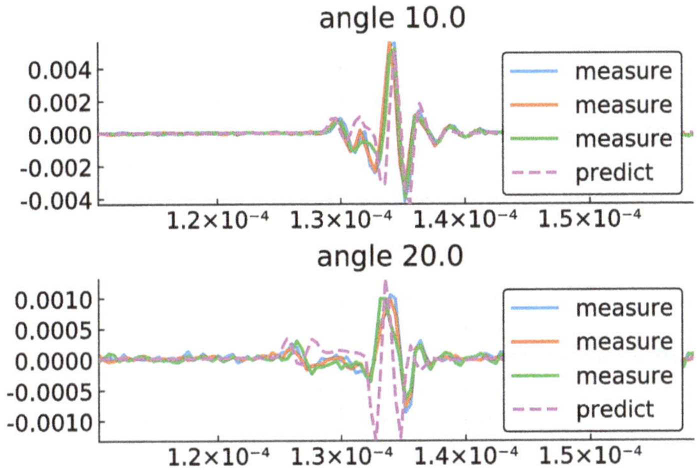

Figure 5.19. Time domain plot of the signal received at (top) 10 o and (bottom) at 20 o for water filling the silicone tube.

5.4 Example 4: liquid diffusion through powders

A detailed description of the theory underpinning acoustic microscopy can be found in [9]. In summary and in the case of acoustic powder characterisation it is important to distinguish the extent of wave scattering which can be classified based on comparisons between the length scale l of the object and the acoustic wavelength λ:

When $l \leqslant \lambda$, the interaction results in a scattered inhomogeneity
When $l \sim \lambda$, the scattering is sensitive to shape and size resonances
When $l \geqslant \lambda$, the scattering is insensitive to shape and size resonances

Parameters that must be considered include the wave attenuation, acoustic contrast, resolution, focusing aberration, scanning time and thermal aberration (figures 5.21–5.24).

This work was part of a University of Leeds MSc project by Gabrielle Lankester.

5.5 Interpreting the airborne acoustic scan data from the ultran scanner

5.5.1 Mathematical introduction

Refer to figure 5.25

5.5.2 Assumptions

Whilst any shape of object can be scanned, we ignore the refraction of the sound beam by the sample caused by the angle of incidence of the beam. Software such as Comsol Multiphysics can be used to account for beam refraction which will tend to increase the path length of the sound beam.

The speed of sound in air does not change during the scan.

The speed of sound in the sample does not change during the scan

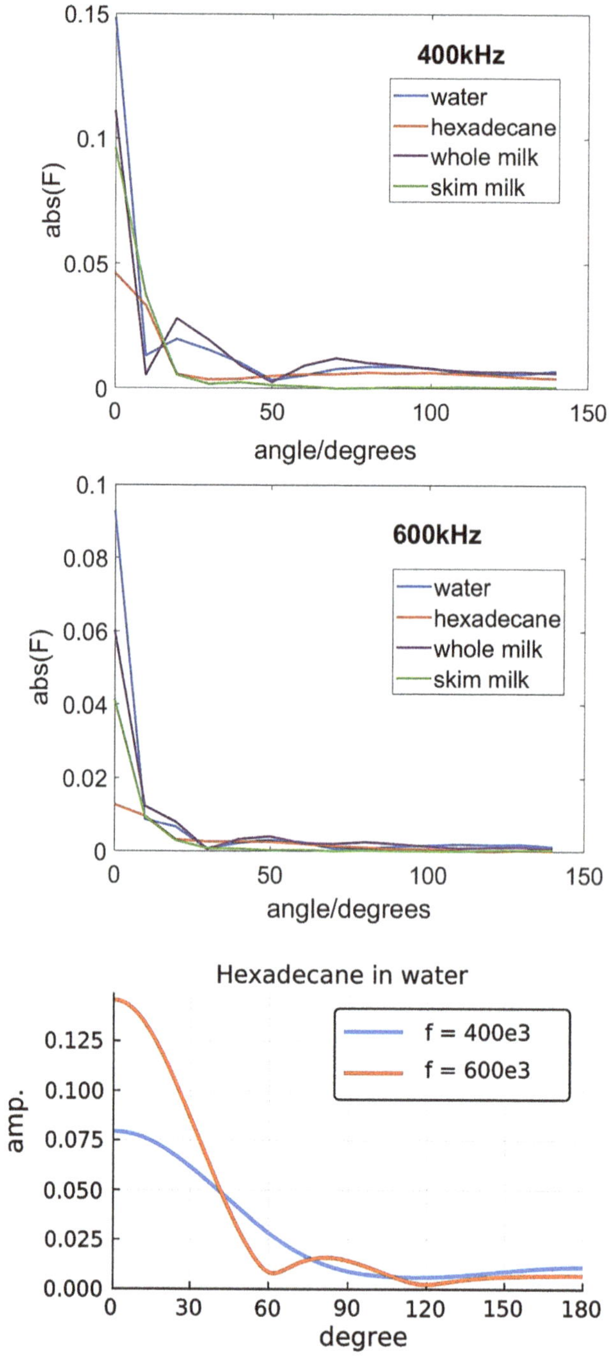

Figure 5.20. Comparison of the angular dependence of the scattered signal in in water, hexadecane, whole and skimmed milk (top) at 400 kHz, (middle) at 600 kHz and (bottom) in an emulsion of hexadecane in water at 400 and 600 kHz. Measurements are courtesy of Matt Francis.

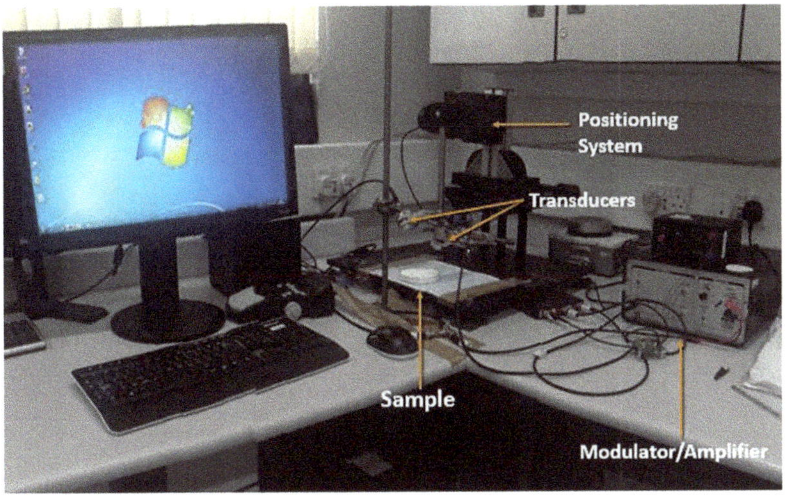

Figure 5.21. Airborne acoustic microscope.

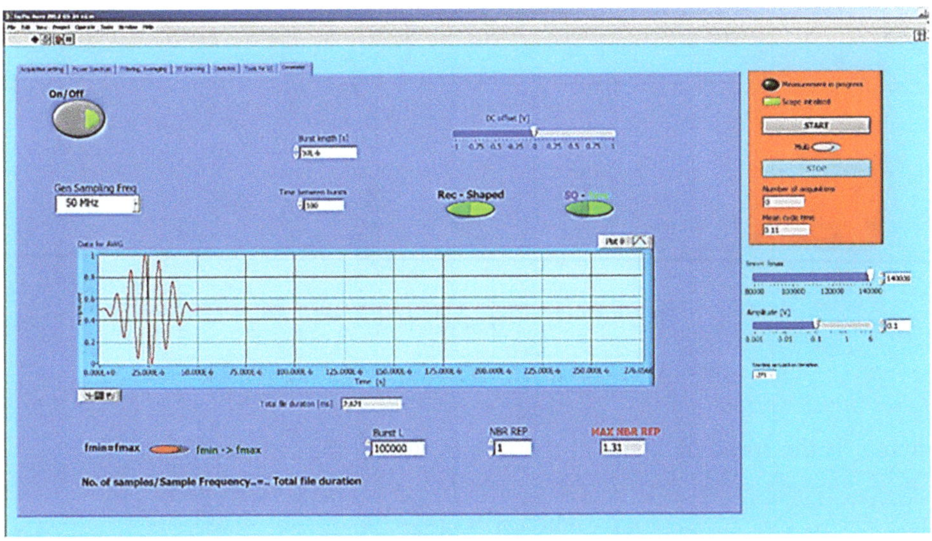

Figure 5.22. Tie Pie Aero user interface showing detected acoustic signal.

If $d_{\text{sample }-i}$ is the ith sample thickness and Δt_i the ith time delay due to the insertion of the sample into the acoustic beam with transducer separation d; Δt_{air} is the time-of-flight in the absence of the sample and Δt_{i} the time-of-flight with the sample inserted, keeping the transducer separation d constant at all positions 1 through i to n then

$$c_i = \frac{d_{\text{sample}-i}}{\Delta t_i - \quad \Delta t_{\text{air}}} \tag{5.4}$$

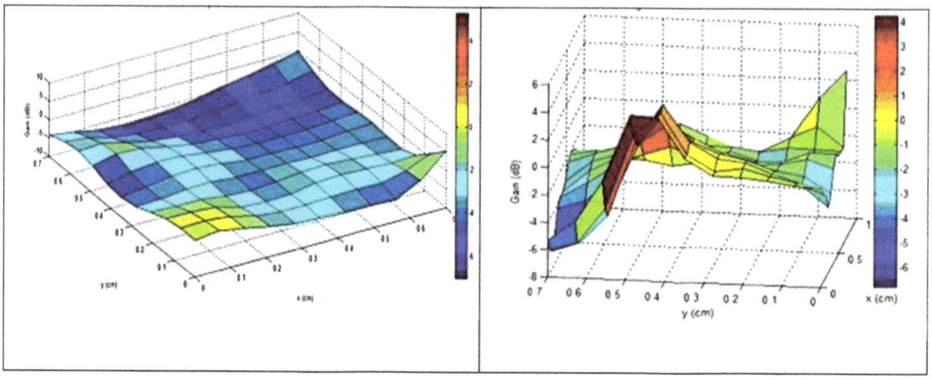

Figure 5.23. Scan of dry powder (left) and powder with a drop of fluid added (right).

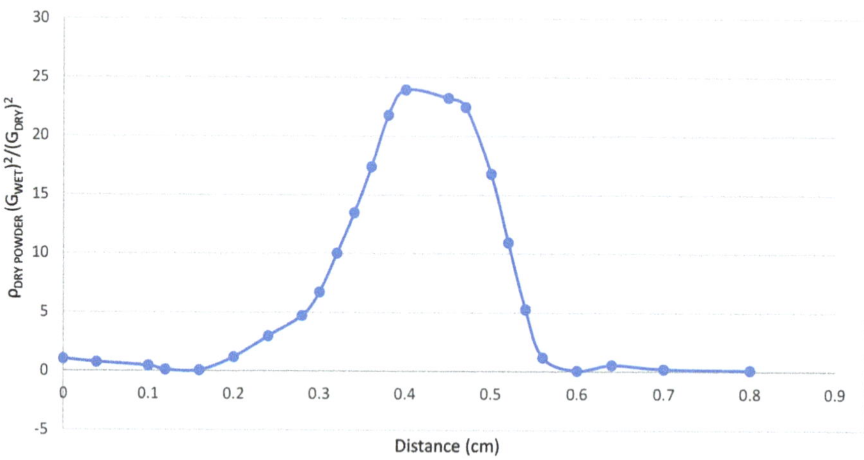

Figure 5.24. Variation of wet powder density with distance.

where c_i is the speed of sound at the *i*th position in the sample.

$$\text{Then } c_i = \sqrt{\frac{M_i}{\rho_i}} \tag{5.5}$$

where

$$M_i = K_i + 4/3G_i \tag{5.6}$$

And M_i is the elastic constant, ρ_i is density, *Ki* is the bulk modulus and *Gi* is the shear modulus.

5.5.3 Solution 1: a homogeneous elastic solid

If we know the density of the sample we can determine the elastic modulus at each position, *i*, M_i using equations (5.5) and (5.6).

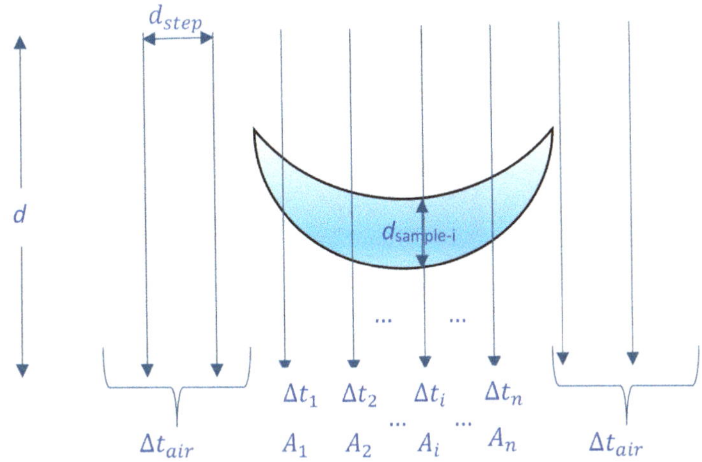

Figure 5.25. The solid curved segment represents a sample resting between supports holding it within the scan area. The scanner steps on a grid size $d_{\text{step}} \cdot d_{\text{step}}$ and the probe separation is a constant d. We assume that the speed of sound in air does not change during the scan. The ith segment of the sample scanned takes time Δt_i, thickness $\text{d}_{\text{sample}-i}$ and the signal amplitude is A_i. Note that this treatment does not account for refraction of the sound due to sample curvature.

5.5.4 Solution 2: inhomogeneous elastic solid with speed of sound less than that in air

That the speed of sound in the sample is less than that in air implies that there is considerable voidage in the sample. In this case we use a simple mixing rule called the Urick equation

$$K = 1/\kappa \tag{5.7}$$

where κ is the adiabatic compressibility and

$$c_{\text{sample}} = \sqrt{\frac{1}{\kappa\rho}} \tag{5.8}$$

where

$$\kappa = \kappa_{\text{void}}\varphi_{\text{void}} + \kappa_{\text{continuous-phase}}(1 - \varphi_{\text{void}}) \tag{5.9}$$

$$\rho = \rho_{\text{void}}\varphi_{\text{void}} + \rho_{\text{continuous-phase}}(1 - \varphi_{\text{void}}) \tag{5.10}$$

Assuming that $K >> G$ then

$$c_{\text{void}} = \sqrt{K_{\text{void}}/\rho_{\text{void}}} = \sqrt{1/\kappa_{\text{void}}\rho_{\text{void}}} \tag{5.11}$$

$$\begin{aligned} c_{\text{continuous-phase}} &= \sqrt{K_{\text{continuous-phase}}/\rho_{\text{continuous-phase}}} \\ &= \sqrt{1/\kappa_{\text{continuous-phase}}\rho_{\text{continuous-phase}}} \end{aligned} \tag{5.12}$$

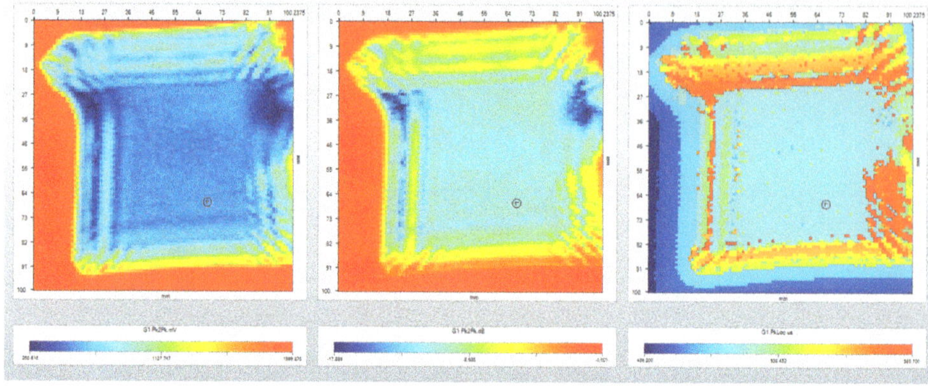

Figure 5.26. Left: Plot of signal amplitude (mV); Middle: Plot of signal amplitude (dB); Right: Plot of signal delay (μs), note that the signal delay in the sample is longer than that in the surrounding air.

These equations can then be solved for φ_{void} to give the voidage as a proportion of the sample at the measured data point.

Example of a sample made from compressed cellulose fibres (figure 5.26)

The speed of sound

$$c = \frac{d_{\text{sample}}}{t_{\text{air+sample}} - t_{\text{air}}}$$

If the time difference is 500 μs and the sample thickness is 3 mm then the speed of sound

$$c = \frac{3 \times 10^{-3}}{500 \times 10^{-6}} = 6 \quad \text{ms}^{-1}$$

$$c = \sqrt{\frac{M}{\rho}}$$

where M is the elastic modulus and ρ is the density of cellulose composite

$$M = \rho c^2$$

$$M = 0.91 \times \frac{10^{-3}}{(10^{-2})^3} \text{kg m}^{-3} \times 0.6^2 \quad \text{m}^{-2} \quad \text{s}^{-2}$$

$$M = 328 \text{ N m}^{-2} \text{ or Pa}$$

$$M = 328 \text{ Pa}$$

The above calculation determines the effective elastic modulus at one point in the scan. However, the sample comprises cellulose fibers whose bulk modulus is 20 GPa and Young's modulus is 15 GPa (https://core.ac.uk/download/pdf/1897844.pdf) and density is 1000 $kg\,m^{-3}$. If we assume that the speed of sound in the voids is close to that of the speed of sound in air, 340 $m\,s^{-1}$ and density 1.29 $kg\,m^{-3}$ then the adiabatic compressibility of the voids will be 1/(340 × 340 × 1.29) from equation (5.12), 6.71 MPa^{-1}.

We measure the speed of sound from the time delay in the sample and from equation (5.9)

$$\frac{\kappa - \kappa_{\text{continuous-phase}}}{\kappa_{\text{void}} - \kappa_{\text{continuous-phase}}} = \varphi_{\text{void}}$$

$$\frac{1/c_{\text{sample}}^{2}\,\rho - \kappa_{\text{continuous-phase}}}{\kappa_{\text{void}} - \kappa_{\text{continuous-phase}}} = \varphi_{\text{void}} \tag{5.13}$$

In the example above the velocity of sound is 6 $m\,s^{-1}$, κ_{void} is 6.71 MPa^{-1}, density is 0.91 $kg\,m^{-3}$ and $\kappa_{\text{continuous-phase}}$ is 0.05 GPa^{-1} which gives the unphysical result for the voidage of 4.5 (it should be less than 1 and greater than 0). However, this can be scaled to get a map of voidage in the sample on the scale of 1 mm × 1 mm.

5.6 Fat crystallisation monitoring using supervised-machine learning

Metilli *et al* [2] use the device shown in figure 5.2 to monitor fat crystallisation in oleogels with the results shown in figures 5.27 and 5.28.

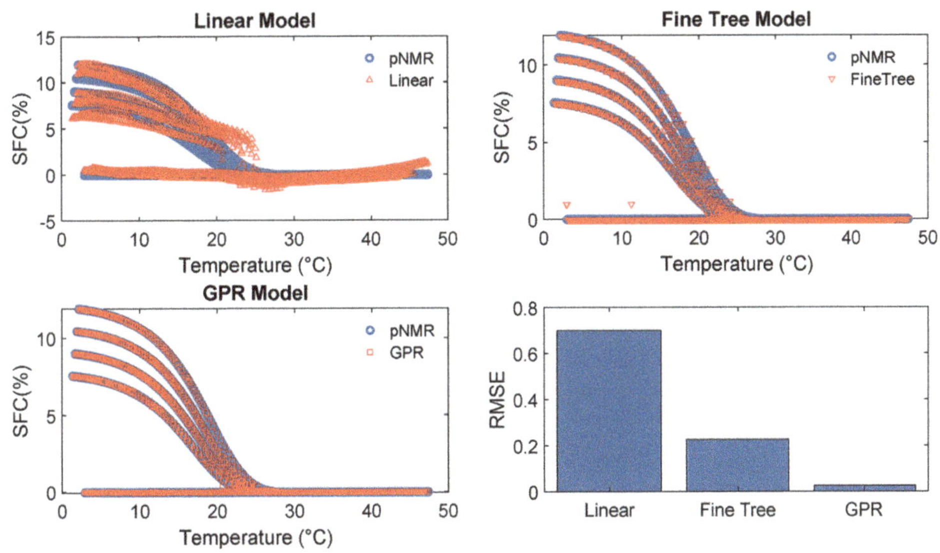

Figure 5.27. Predicted SFC(%) values obtained from the supervised-machine learning models, compared with the measured SFC(%) from pNMR (blue circles) as a function of temperature: (a) Linear model (red upward triangles), (b) Fine Tree model (red downward triangles), (c) Gaussian Process Regression (red squares) and (d) histogram plot showing the RMSE on the SFC(%) values for each model.

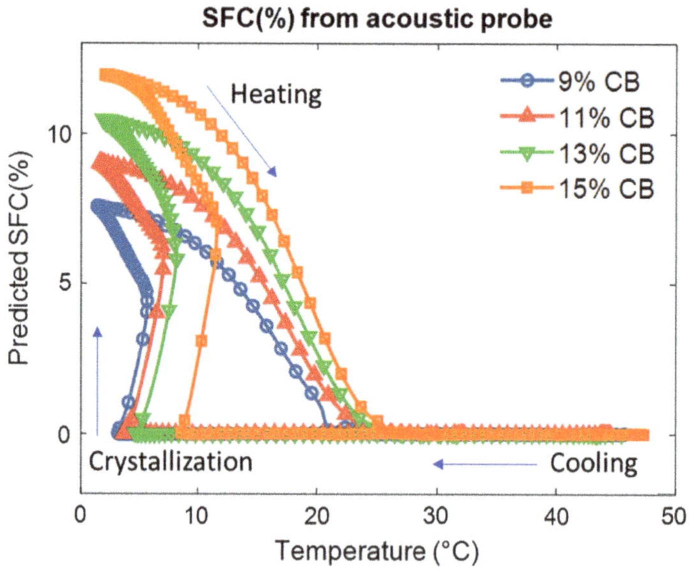

Figure 5.28. Evolution of ultrasound predicted SFC% during cooling from 40 °C 5 °C and heating (5 °C–40 °C to C) for oleogels containing different CB w/w%.

References

[1] Cao Y, Huang C, Guo Y, Xu Y, Gong S, Chu Q and Chen P 2024 Unraveling the contributing factors of stale odor in Longjing tea through a sensomics approach *Food Chem.* **441** 138301

[2] Metilli L, Morris L, Lazidis A, Marty-Terrade S, Holmes M, Povey M and Simone E 2022 Real-time monitoring of fat crystallization using pulsed acoustic spectroscopy and supervised machine learning *J. Food Eng.* **335** 111192

[3] Povey M, Chen J, Karlsson C, Bank J and Simonin B 2005 Acoustic detection of crack propagation and characterisation of material properties therebye *AFPAC* **2005** 13

[4] Varela P, Chen J, Fiszman S and Povey M J W 2006 Crispness assessment of roasted almonds by an integrated approach to texture description: texture, acoustics, sensory and structure *J. Chemom.* **20** 311–20

[5] Chen J S, Karlsson C and Povey M 2005 Acoustic envelope detector for crispness assessment of biscuits *J. Texture Stud.* **36** 139–56

[6] Gregersen S B, Povey M J W, Andersen M D, Hammershøj M, Rappolt M, Sadeghpour A and Wiking L 2016 Acoustic properties of crystallized fat: relation between polymorphic form, microstructure, fracturing behavior, and sound intensity *Eur. J. Lipid Sci. Technol.* **118** 1257–70

[7] Gregersen S B, Povey M J W, Kidmose U, Andersen M D, Hammershoj M and Wiking L 2015 Identification of important mechanical and acoustic parameters for the sensory quality of cocoa butter alternatives *Food Res. Int.* **76** 637–44

[8] Hefft D I, Povey M J and Zhang Z J 2025 Transient energy released by a tribological process *Tribol. Int.* **212** 110928

[9] Povey M J W, Watson N and Parker N G 2013 Ultrasonic and acoustic microscopy: principles and applications to food microstructures *Food Microstructures* ed V J M Groves (Woodhead Publishing) 192–222

IOP Publishing

Ultrasonic Techniques for Liquids Characterisation

Megan J Povey

Chapter 6

The future?

Megan J Povey

When CoPilot was asked this question, it provided the following answer.

The future of ultrasound (ultrasonic) sensors across the **food, aerospace, pharmaceutical, agrochemical, and automotive industries** is promising, driven by technological innovation, automation, and the demand for precision and safety. Here we provide a breakdown by sector.

6.1 Food industry

Ultrasound sensors are increasingly used for:

- **Non-destructive testing (NDT)** of food texture, composition, and quality.
- **Level and flow measurement** in liquid food processing.
- **Cleaning and sterilization** using high-frequency ultrasound.
- **Detection of foreign bodies** in packaged goods.
- Cleaning in place - detection of fouling.
- Food materials characterization and objective measures of human sensory attributes such as crispy/crunchiness.

The trend is toward **sustainable, hygienic, and automated processing**, with ultrasound playing a key role in real-time monitoring and quality assurance.

6.2 Aerospace industry

While not as widely adopted as in other sectors, ultrasound sensors are gaining traction for:

- **Structural health monitoring** (e.g., detecting cracks or corrosion in aircraft components).
- **Non-invasive inspection** of composite materials.
- **Fluid level sensing** in fuel and hydraulic systems.

doi:10.1088/978-0-7503-3203-3ch6

Future developments may include **integration with AI** for predictive maintenance and **miniaturized sensors** for UAVs and satellites.

6.3 Pharmaceutical industry

Ultrasound sensors support:

- **Precise liquid level detection** in sterile environments.
- **Monitoring of mixing and crystallization** processes.
- **Non-contact measurement** in cleanrooms to avoid contamination.

As **personalized medicine and continuous manufacturing** grow, ultrasound will be vital for **real-time process analytics** and **regulatory compliance**.

6.4 Agrochemical industry

Applications include:

- **Tank level monitoring** for fertilizers and pesticides.
- **Flow control** in irrigation and spraying systems.
- **Soil and crop analysis** using ultrasonic imaging.

The push toward **precision agriculture** and **environmental sustainability** is driving adoption, especially when paired with IoT and drone technologies.

6.5 Automotive industry

This is one of the most mature sectors for ultrasound sensors, with uses such as:

- **Parking assistance and obstacle detection** (ADAS).
- **Blind spot monitoring** and **autonomous navigation**.
- **Fluid level sensing** (e.g., oil, coolant, washer fluid).

The future lies in **autonomous vehicles**, where ultrasound complements radar and LiDAR for **close-range sensing** and **safety systems** [1].

6.6 Cross-industry trends

- **Integration with AI and IoT** for smart sensing and predictive analytics. See chapter 4
- **Miniaturization and energy efficiency** for embedded and wearable applications.
- **Industry 4.0 and automation** driving demand for real-time, non-contact sensing.
- **Sustainability and safety** as key motivators for adoption.

Reference

[1] Ultrasonic Sensors Market (2025–2030) https://www.grandviewresearch.com/industry-analysis/ultrasound-sensors-market (Accessed 4/11/2025)

Chapter 7

Notation

Megan J Povey

a	Particle radius
A_n, B_n, C_n	Coefficients of the scattered wave potentials for compressional, thermal and shear waves respectively
c	(Adiabatic) wave speed of longitudinal mode
$C_{p,v}$	Specific heat capacity at constant pressure or constant volume respectively
k_c, k_t, k_s	Wavenumbers of compressional, thermal and shear wave modes respectively
K_C	Effective wavenumber of the compressional wave in inhomogeneous medium
K_T, K_S	Elastic bulk modulus in isothermal and adiabatic conditions respectively
n_ϕ	Number density of particles (number per unit volume) in system
p	(Thermodynamic) pressure variation wave mode (C,T,S)
t	Time
T	Temperature variation
T_0	Background temperature
$\mathbf{v}$	Material velocity
α	Attenuation
β	Coefficient of volumetric expansion (thermal expansivity)
φ	Scalar wave potential
ε	Strain
$\gamma = C_p/C_v$	Ratio of specific heat capacities
Γ_c	Thermal factor, relating temperature to wave potential
η_μ	Shear viscosity
η_B	Bulk viscosity
κ	Thermal conductivity
λ_c, λ_t, λ_s	Wavelengths of compressional, thermal and shear wave modes respectively
μ	Shear modulus
ρ	Density variation

doi:10.1088/978-0-7503-3203-3ch7

(*Continued*)

ρ_{-}	Density in the unperturbed medium
σ	Stress
τ	Thermal diffusivity $\tau = \dfrac{\kappa}{\rho_0 C_p}$
ω	Angular frequency
Φ_0	Magnitude of incident wave velocity potential
$\boldsymbol{\xi}$	Material displacement
$\boldsymbol{\psi}$	Vector wave potential
Ω	Vorticity

IOP Publishing

Ultrasonic Techniques for Liquids Characterisation

Megan J Povey

Chapter 8

Appendix

Megan J Povey

Errata in the 1997 edition of *Ultrasonic Techniques for Fluids Characterization* [1].

28—Figure 2.17 has incorrect data. The speed of sound in pure water at 20 °C is 1365, not 1500 m s^{-1}

37—Equation (2.2) should read

$$V_{M_1} = V_M - \frac{dV_M}{dc_M} c_M$$

49—Para 4, line 10—right bracket omitted

50—Equation (3.1) is incorrect. It should read $2.3177 - 0.06230T - 0.0038v - 1.0274 \times 10^{-5}T^2 + 1.5208 \times 10^{-6}v^2 + 3.8731 \times 10^{-5}vT$

59—Reference to Coupland missing [2].

63—Hildebrand (equation (3.9)) is incorrect. It should read

$$\ln x = \Delta H_f R\left(\frac{1}{T} - \frac{1}{T_{mp}}\right)$$

106—Dots should replace *X*'s in equations (4.9)–(4.11) for the scalar products

107—Equations (4.12) and (4.14) should read.

$$\dot{\xi} = -\nabla\varphi$$

$$S = Tr\left(S'_{ij}\right) = S'_{ij} = \frac{\partial \xi_i}{\partial x_i} = -\nabla^2\varphi$$

doi:10.1088/978-0-7503-3203-3ch8

108—Equation (4.22): Right braces should join the wave vectors on lines one above the other to the equals sign in between.

114—Equation (4.33): The Gamma and the *k* should be subscripted *c*, *t* referring to the compressional and thermal modes.

116—Equation (4.40): This definition of the total attenuation is incorrect. The scattering contribution α_s is normally computed as the excess attenuation and therefore the contribution from the dispersed phase is implicit:

$$\alpha_{\text{total}} = \alpha_s + (1 - \phi)\alpha_1.$$

References

[1] Povey M J W 1997 *Ultrasonic Techniques for Fluids Characterization* (San Diego, CA: Academic) p 214

[2] Coupland J, Dickinson E, McClements D J, Povey M J W and de Rancourt de Mimmerand C 1993 Crystallization in simple paraffins and monoacid saturated triacylglycerols dispersed in water *Food Colloids and Polymers: Stability and Mechanical Properties* ed E Dickinson and P Walstra (Cambridge: Royal Society of Chemistry) pp 243–9

www.ingramcontent.com/pod-product-compliance
Ingram Content Group UK Ltd.
Pitfield, Milton Keynes, MK11 3LW, UK
UKHW051848100926
8556IPUK00005B/515

9 780750 332019